Heating, Ventilation and Air Conditioning: Design, Analysis and Control Systems

Heating, Ventilation and Air Conditioning: Design, Analysis and Control Systems

Leila Alistair

Larsen & Keller
www.larsen-keller.com

Heating, Ventilation and Air Conditioning: Design, Analysis and Control Systems
Leila Alistair
ISBN: 978-1-64172-123-3 (Hardback)

Larsen & Keller

Published by Larsen and Keller Education,
5 Penn Plaza,
19th Floor,
New York, NY 10001, USA

Cataloging-in-Publication Data

Heating, ventilation and air conditioning : design, analysis and control systems / Leila Alistair.
     p. cm.
Includes bibliographical references and index.
ISBN 978-1-64172-123-3
1. Heating. 2. Ventilation. 3. Air conditioning. 4. Engineering design.
5. Electronic apparatus and appliances--Temperature control. 6. Air conditioning-- Control. I. Alistair, Leila.
TH7222 .H43 2019
697--dc23

For more information regarding Larsen and Keller Education and its products, please visit the publisher's website www.larsen-keller.com

Table of Contents

Permissions

Index

Preface

Heating, ventilation and air conditioning is a technology that is concerned with indoor and vehicular environmental comfort. Its objective is to provide comfort and high indoor air quality. The technology develops on the principles of fluid mechanics, thermodynamics and heat transfer. Ventilation involves exchanging air in any space in order to control temperature as well as remove odors, dust, airborne bacteria, carbon dioxide, etc. It can be achieved mechanically by using an air handler, mechanical exhausts or ceiling fans, or naturally using operable windows, louvers or trickle vents. In central heating, water, steam or air is heated using a boiler, furnace or heat pump, and the resultant heat is transferred by the processes of convection, radiation or conduction to the living spaces in a house or building. Air conditioning and refrigeration involves cooling and humidity control through the removal of heat using heat transfer processes. This book is a compilation of chapters that discuss the most vital concepts about the technology of heating, ventilation and air conditioning. Such selected concepts that redefine the understanding of the crucial aspects of this technology including its design, analysis and control systems have been presented herein. It will serve as a valuable reference guide for architects, interior designers, professionals and students involved in this area of study.

A foreword of all chapters of the book is provided below:

Chapter 1, HVAC also known as heating, ventilation and air conditioning, is the technology that is meant to provide a comfortable and acceptable air quality in indoor settings and vehicles. The topics elaborated in this chapter on thermal comfort, thermal destratification, thermal mass, Stack effect, sensible heat, etc. will help in providing a better perspective about the fundamental concepts of HVAC; **Chapter 2**, Ventilation is the introduction of ambient air into a space for the control of indoor air quality by diluting and displacing pollutants from the indoor air setting. This chapter discusses in detail the concepts and principles of airflow, air infiltration and exfiltration, air distribution, displacement ventilation, etc. for an extensive understanding of ventilation and air distribution; **Chapter 3**, A central heating system is used to provide heat and warmth to the interiors of a building from one point to multiple rooms. All the diverse aspects of central heating systems such as geothermal heat pump, air source heat pump, ground-coupled heat exchanger, space heater, underfloor heating, etc. have been carefully analyzed in this chapter; **Chapter 4**, Air conditioning is the technology of removing moisture and heat from an interior space for the comfort of the occupants. This chapter has been carefully written to provide an easy understanding of the varied aspects of air conditioning, such as district cooling system, chilled beam system, evaporative cooler, deep water source cooling, etc.; **Chapter 5**, Science and technology have undergone rapid developments in the past decade, which has resulted in innovation in HVAC systems. The following chapter elucidates a study of the measurements for HVAC and includes vital topics related to air flow meter, blower door, gas detector, thermostat, carbon dioxide sensor, etc.; **Chapter 6**, In order to control or regulate the heating or air conditioning, we require HVAC control systems. This chapter includes fundamental topics related to building automation, direct digital control systems, electrical and electronic control systems, microprocessor systems, etc. for a complete understanding of HVAC control systems.

At the end, I would like to thank all the people associated with this book devoting their precious time and providing their valuable contributions to this book. I would also like to express my gratitude to my fellow colleagues who encouraged me throughout the process.

Leila Alistair

Fundamental Concepts of HVAC

HVAC also known as heating, ventilation and air conditioning, is the technology that is meant to provide a comfortable and acceptable air quality in indoor settings and vehicles. The topics elaborated in this chapter on thermal comfort, thermal destratification, thermal mass, Stack effect, sensible heat, etc. will help in providing a better perspective about the fundamental concepts of HVAC.

HVAC stands for heating, ventilation and air conditioning (HVAC). It refers to the different systems, machines and technologies used in indoor settings such as homes, offices and hallways, and transportation systems that need environmental regulation to improve comfort.

For example, HVAC technicians would be hired to create the ideal environment in a subway or bus. They would also be hired to repair and install heating and air-conditioning systems in a new home that was being built, or to regulate the humidity level of an office.

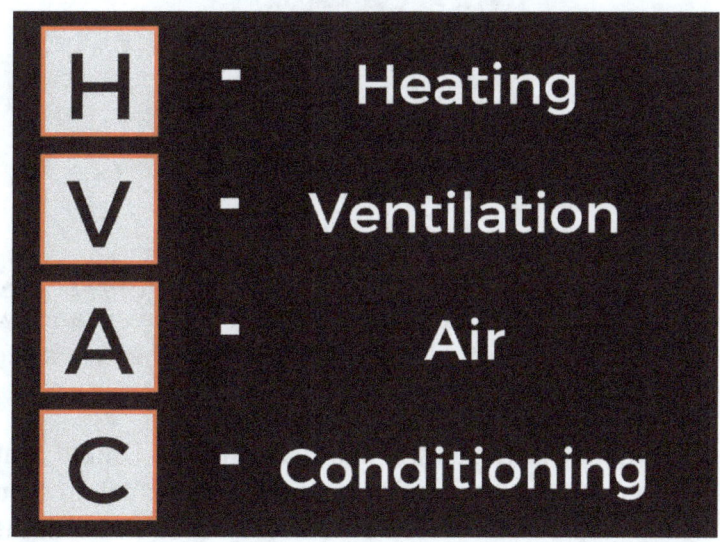

The general HVAC meaning can be defined as a system that provides different types of heating and cooling services to residential and commercial buildings, and for various types of vehicles according to industry standards.

The primary function of the HVAC system in your home – or the heating, ventilation and air condition system – is to provide thermal comfort (control the temperature) and produce acceptable indoor air quality (by controling humidity and filtering the air). But, how does it work?

Understanding the basic functions of an HVAC system and how it works is important to maintaining your current system in good condition.

Heating

Heaters are appliances whose purpose is to generate heat (i.e. warmth) for the building. This can be done via central heating. Such a system contains a boiler, furnace, or heat pump to heat water, steam, or air in a central location such as a furnace room in a home, or a mechanical room in a large building. The heat can be transferred by convection, conduction, or radiation.

Generation

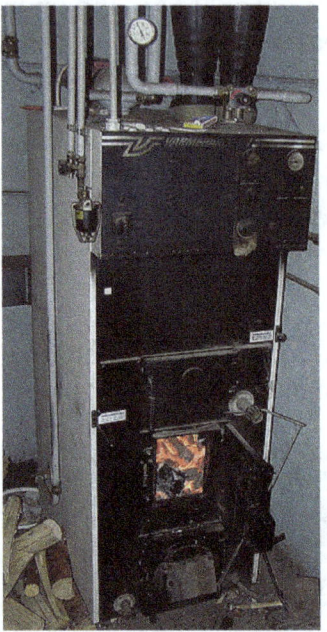

Central heating unit

Heaters exist for various types of fuel, including solid fuels, liquids, and gases. Another type of heat source is electricity, normally heating ribbons composed of high resistance wire. This principle is also used for baseboard heaters and portable heaters. Electrical heaters are often used as backup or supplemental heat for heat pump systems.

The heat pump gained popularity in the 1950s in Japan and the United States. Heat pumps can extract heat from various sources, such as environmental air, exhaust air from a building, or from the ground. Initially, heat pump HVAC systems were only used in moderate climates, but with improvements in low temperature operation and reduced loads due to more efficient homes, they are increasing in popularity in cooler climates.

Distribution

Water/steam

In the case of heated water or steam, piping is used to transport the heat to the rooms. Most modern hot water boiler heating systems have a circulator, which is a pump, to move hot water through the distribution system (as opposed to older gravity-fed systems). The heat can be transferred to the surrounding air using radiators, hot water coils (hydro-air), or other heat exchangers. The radiators may be mounted on walls or installed within the floor to produce floor heat.

The use of water as the heat transfer medium is known as hydronics. The heated water can also supply an auxiliary heat exchanger to supply hot water for bathing and washing.

Air

Warm air systems distribute heated air through duct work systems of supply and return air through metal or fiberglass ducts. Many systems use the same ducts to distribute air cooled by an evaporator coil for air conditioning. The air supply is normally filtered through air cleaners to remove dust and pollen particles.

Dangers

The use of furnaces, space heaters, and boilers as a method of indoor heating could result in incomplete combustion and the emission of carbon monoxide, nitrogen oxides, formaldehyde, volatile organic compounds, and other combustion byproducts. Incomplete combustion occurs when there is insufficient oxygen; the inputs are fuels containing various contaminants and the outputs are harmful byproducts, most dangerously carbon monoxide, which is a tasteless and odorless gas with serious adverse health effects.

Without proper ventilation, carbon monoxide can be lethal at concentrations of 1000 ppm (0.1%). However, at several hundred ppm, carbon monoxide exposure induces headaches, fatigue, nausea, and vomiting. Carbon monoxide binds with hemoglobin in the blood, forming carboxyhemoglobin, reducing the blood's ability to transport oxygen. The primary health concerns associated with carbon monoxide exposure are its cardiovascular and neurobehavioral effects. Carbon monoxide can cause atherosclerosis (the hardening of arteries) and can also trigger heart attacks. Neurologically, carbon monoxide exposure reduces hand to eye coordination, vigilance, and continuous performance. It can also affect time discrimination.

Ventilation

Ventilation is the process of changing or replacing air in any space to control temperature or remove any combination of moisture, odors, smoke, heat, dust, airborne bacteria, or carbon dioxide, and to replenish oxygen. Ventilation includes both the exchange of air with the outside as well as circulation of air within the building. It is one of the most important factors for maintaining acceptable indoor air quality in buildings. Methods for ventilating a building may be divided into *mechanical/forced* and *natural* types.

Mechanical or Forced Ventilation

HVAC ventilation exhaust for a 12-story building

Mechanical, or forced, ventilation is provided by an air handler (AHU) and used to control indoor air quality. Excess humidity, odors, and contaminants can often be controlled via dilution or replacement with outside air. However, in humid climates more energy is required to remove excess moisture from ventilation air.

Kitchens and bathrooms typically have mechanical exhausts to control odors and sometimes humidity. Factors in the design of such systems include the flow rate (which is a function of the fan speed and exhaust vent size) and noise level. Direct drive fans are available for many applications, and can reduce maintenance needs.

Ceiling fans and table/floor fans circulate air within a room for the purpose of reducing the perceived temperature by increasing evaporation of perspiration on the skin of the occupants. Because hot air rises, ceiling fans may be used to keep a room warmer in the winter by circulating the warm stratified air from the ceiling to the floor.

Natural Ventilation

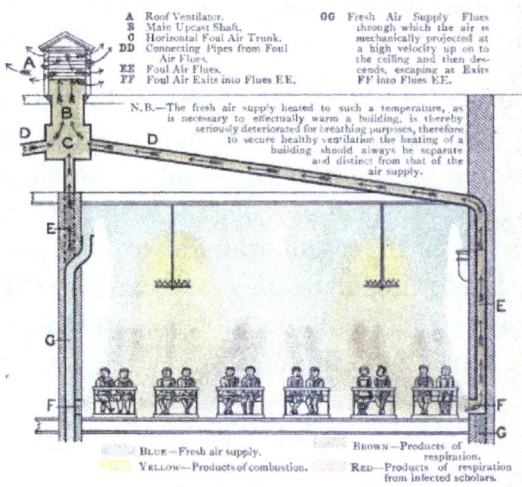

Ventilation on the downdraught system, by impulsion, or the 'plenum' principle, applied to schoolrooms

Natural ventilation is the ventilation of a building with outside air without using fans or other mechanical systems. It can be via operable windows, louvers, or trickle vents when spaces are small and the architecture permits. In more complex schemes, warm air is allowed to rise and flow out high building openings to the outside (stack effect), causing cool outside air to be drawn into low building openings. Natural ventilation schemes can use very little energy, but care must be taken to ensure comfort. In warm or humid climates, maintaining thermal comfort solely via natural ventilation might not be possible. Air conditioning systems are used, either as backups or supplements. Air-side economizers also use outside air to condition spaces, but do so using fans, ducts, dampers, and control systems to introduce and distribute cool outdoor air when appropriate.

An important component of natural ventilation is air change rate or air changes per hour: the hourly rate of ventilation divided by the volume of the space. For example, six air changes per hour means an amount of new air, equal to the volume of the space, is added every ten minutes. For human comfort, a minimum of four air changes per hour is typical, though warehouses might have only two. Too high of an air change rate may be uncomfortable, akin to a wind tunnel which have thousands of changes per hour. The highest air change rates are for crowded spaces, bars, night clubs, commercial kitchens at around 30 to 50 air changes per hour.

Room pressure can be either positive or negative with respect to outside the room. Positive pressure occurs when there is more air being supplied than exhausted, and is common to reduce the infiltration of outside contaminants.

Airborne Diseases

Natural ventilation is a key factor in reducing the spread of airborne illnesses such as tuberculosis, the common cold, influenza and meningitis. Opening doors, windows, and using ceiling fans are all ways to maximize natural ventilation and reduce the risk of airborne contagion. Natural ventilation requires little maintenance and is inexpensive.

Air Conditioning

An air conditioning system, or a standalone air conditioner, provides cooling and humidity control for all or part of a building. Air conditioned buildings often have sealed windows, because open windows would work against the system intended to maintain constant indoor air conditions. Outside, fresh air is generally drawn into the system by a vent into the indoor heat exchanger section, creating positive air pressure. The percentage of return air made up of fresh air can usually be manipulated by adjusting the opening of this vent. Typical fresh air intake is about 10%.

Air conditioning and refrigeration are provided through the removal of heat. Heat can be removed through radiation, convection, or conduction. Refrigeration conduction media such as water, air, ice, and chemicals are referred to as refrigerants. A refrigerant is employed either in a heat pump system in which a compressor is used to drive thermodynamic refrigeration cycle, or in a free cooling system which uses pumps to circulate a cool refrigerant (typically water or a glycol mix).

Refrigeration Cycle

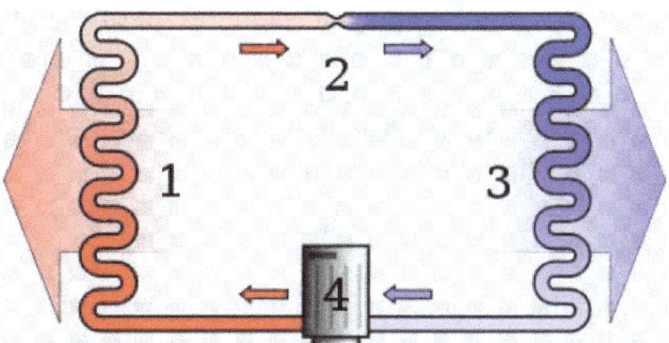

A simple stylized diagram of the refrigeration cycle: 1) condensing coil,
2) expansion valve, 3) evaporator coil, 4) compressor

The refrigeration cycle uses four essential elements to cool.

- The system refrigerant starts its cycle in a gaseous state. The compressor pumps the refrigerant gas up to a high pressure and temperature.

- From there it enters a heat exchanger (sometimes called a condensing coil or condenser) where it loses energy (heat) to the outside, cools, and condenses into its liquid phase.

- An expansion valve (also called metering device) regulates the refrigerant liquid to flow at the proper rate.

- The liquid refrigerant is returned to another heat exchanger where it is allowed to evaporate, hence the heat exchanger is often called an evaporating coil or evaporator. As the liquid refrigerant evaporates it absorbs energy (heat) from the inside air, returns to the compressor, and repeats the cycle. In the process, heat is absorbed from indoors and transferred outdoors, resulting in cooling of the building.

In variable climates, the system may include a reversing valve that switches from heating in winter to cooling in summer. By reversing the flow of refrigerant, the heat pump refrigeration cycle is changed from cooling to heating or vice versa. This allows a facility to be heated and cooled by a single piece of equipment by the same means, and with the same hardware.

Free Cooling

Free cooling systems can have very high efficiencies, and are sometimes combined with seasonal thermal energy storage so that the cold of winter can be used for summer air conditioning. Common storage mediums are deep aquifers or a natural underground rock mass accessed via a cluster of small-diameter, heat-exchanger-equipped boreholes. Some systems with small storages are hybrids, using free cooling early in the cooling season, and later employing a heat pump to chill the circulation coming from the storage. The heat pump is added-in because the storage acts as a heat sink when the system is in cooling (as opposed to charging) mode, causing the temperature to gradually increase during the cooling season.

Some systems include an "economizer mode", which is sometimes called a "free-cooling mode". When economizing, the control system will open (fully or partially) the outside air damper and close (fully or partially) the return air damper. This will cause fresh, outside air to be supplied to the system. When the outside air is cooler than the demanded cool air, this will allow the demand to be met without using the mechanical supply of cooling (typically chilled water or a direct expansion "DX" unit), thus saving energy. The control system can compare the temperature of the outside air vs. return air, or it can compare the enthalpy of the air, as is frequently done in climates where humidity is more of an issue. In both cases, the outside air must be less energetic than the return air for the system to enter the economizer mode.

Packaged vs. Split System

Central, "all-air" air-conditioning systems (or package systems) with a combined outdoor condenser/evaporator unit are often installed in North American residences, offices, and public buildings, but are difficult to retrofit (install in a building that was not designed to receive it) because of the bulky air ducts required. (Minisplit ductless systems are used in these situations.) Outside of North America, packaged systems are only used in limited applications involving large indoor space such as stadiums, theatres or exhibition halls.

An alternative to packaged systems is the use of separate indoor and outdoor coils in split systems.

Split systems are preferred and widely used worldwide except in the North America. In the North America, split systems are most often seen in residential applications, but they are gaining popularity in small commercial buildings.

With the split system, the evaporator coil is connected to a remote condenser unit using refrigerant piping between an indoor and outdoor unit instead of ducting air directly from the outdoor unit. Indoor units with directional vents mount onto walls, suspended from ceilings, or fit into the ceiling. Other indoor units mount inside the ceiling cavity, so that short lengths of duct handle air from the indoor unit to vents or diffusers around the rooms.

Split systems are more efficient and the footprint is typically smaller than the package systems. On the other hand, package systems tend to have slightly lower indoor noise level compared to split system since the fan motor is located outside.

Dehumidification

Dehumidification (air drying) in an air conditioning system is provided by the evaporator. Since the evaporator operates at a temperature below the dew point, moisture in the air condenses on the evaporator coil tubes. This moisture is collected at the bottom of the evaporator in a pan and removed by piping to a central drain or onto the ground outside.

A dehumidifier is an air-conditioner-like device that controls the humidity of a room or building. It is often employed in basements which have a higher relative humidity because of their lower temperature (and propensity for damp floors and walls). In food retailing establishments, large open chiller cabinets are highly effective at dehumidifying the internal air. Conversely, a humidifier increases the humidity of a building.

Maintenance

All modern air conditioning systems, even small window package units, are equipped with internal air filters. These are generally of a lightweight gauzy material, and must be replaced or washed as conditions warrant. For example, a building in a high dust environment, or a home with furry pets, will need to have the filters changed more often than buildings without these dirt loads. Failure to replace these filters as needed will contribute to a lower heat exchange rate, resulting in wasted energy, shortened equipment life, and higher energy bills; low air flow can result in iced-over evaporator coils, which can completely stop air flow. Additionally, very dirty or plugged filters can cause overheating during a heating cycle, and can result in damage to the system or even fire.

Because an air conditioner moves heat between the indoor coil and the outdoor coil, both must be kept clean. This means that, in addition to replacing the air filter at the evaporator coil, it is also necessary to regularly clean the condenser coil. Failure to keep the condenser clean will eventually result in harm to the compressor, because the condenser coil is responsible for discharging both the indoor heat (as picked up by the evaporator) and the heat generated by the electric motor driving the compressor.

Energy Efficiency

Since the 1980s, manufacturers of HVAC equipment have been making an effort to make the systems they manufacture more efficient. This was originally driven by rising energy costs, and has more recently been driven by increased awareness of environmental issues. Additionally, improvements to the HVAC system efficiency can also help increase occupant health and productivity. In the US, the EPA has imposed tighter restrictions over the years. There are several methods for making HVAC systems more efficient.

Heating Energy

In the past, water heating was more efficient for heating buildings and was the standard in the United States. Today, forced air systems can double for air conditioning and are more popular.

Some benefits of forced air systems, which are now widely used in churches, schools and high-end residences, are:

- Better air conditioning effects

- Energy savings of up to 15-20%

- Even conditioning

A drawback is the installation cost, which can be slightly higher than traditional HVAC systems.

Energy efficiency can be improved even more in central heating systems by introducing zoned heating. This allows a more granular application of heat, similar to non-central heating systems. Zones are controlled by multiple thermostats. In water heating systems the thermostats control zone valves, and in forced air systems they control zone dampers inside the vents which selectively block the flow of air. In this case, the control system is very critical to maintaining a proper temperature.

Forecasting is another method of controlling building heating by calculating demand for heating energy that should be supplied to the building in each time unit.

Ground Source Heat Pump

Ground source, or geothermal, heat pumps are similar to ordinary heat pumps, but instead of transferring heat to or from outside air, they rely on the stable, even temperature of the earth to provide heating and air conditioning. Many regions experience seasonal temperature extremes, which would require large-capacity heating and cooling equipment to heat or cool buildings. For example, a conventional heat pump system used to heat a building in Montana's −70° F (−57° C) low temperature or cool a building in the highest temperature ever recorded in the US—134° F (57° C) in Death Valley, California, in 1913 would require a large amount of energy due to the extreme difference between inside and outside air temperatures. A few feet below the earth's surface, however, the ground remains at a relatively constant temperature. Utilizing this large source of relatively moderate temperature earth, a heating or cooling system's capacity can often be significantly reduced. Although ground temperatures vary according to latitude, at 6 feet (1.8 m) underground, temperatures generally only range from 45 to 75° F (7 to 24° C).

An example of a geothermal heat pump that uses a body of water as the heat sink, is the system used by the Trump International Hotel and Tower in Chicago, Illinois. This building is situated on the Chicago River, and uses cold river water by pumping it into a recirculating cooling system, where heat exchangers transfer heat from the building into the water, and then the now-warmed water is pumped back into the Chicago River.

While they may be more costly to install than regular heat pumps, geothermal heat pumps can produce markedly lower energy bills – 30 to 40 percent lower, according to estimates from the US Environmental Protection Agency.

Geothermal heat pumps still provide higher efficiency than air source heat pumps. Some models provide 70% saving compared to electric resistance heaters. .

Ventilation Energy Recovery

Energy recovery systems sometimes utilize heat recovery ventilation or energy recovery ventilation systems that employ heat exchangers or enthalpy wheels to recover sensible or latent heat from exhausted air. This is done by transfer of energy to the incoming outside fresh air.

Air Conditioning Energy

The performance of vapor compression refrigeration cycles is limited by thermodynamics. These air conditioning and heat pump devices *move* heat rather than convert it from one form to another, so *thermal efficiencies* do not appropriately describe the performance of these devices. The Coefficient-of-Performance (COP) measures performance, but this dimensionless measure has not been adopted. Instead, the Energy Efficiency Ratio (*EER*) has traditionally been used to characterize the performance of many HVAC systems. EER is the Energy Efficiency Ratio based on a 35° C (95° F) outdoor temperature. To more accurately describe the performance of air conditioning equipment over a typical cooling season a modified version of the EER, the Seasonal Energy Efficiency Ratio (*SEER*), or in Europe the ESEER, is used. SEER ratings are based on seasonal temperature averages instead of a constant 35° C (95° F) outdoor temperature. The current industry minimum SEER rating is 14 SEER.

Engineers have pointed out some areas where efficiency of the existing hardware could be improved. For example, the fan blades used to move the air are usually stamped from sheet metal, an economical method of manufacture, but as a result they are not aerodynamically efficient. A well-designed blade could reduce electrical power required to move the air by a third.

Air Filtration and Cleaning

Air handling unit, used for heating, cooling, and filtering the air

Air cleaning and filtration removes particles, contaminants, vapors and gases from the air. The filtered and cleaned air then is used in heating, ventilation and air conditioning. Air cleaning and filtration should be taken in account when protecting our building environments.

Clean Air Delivery Rate and Filter Performance

Clean air delivery rate is the amount of clean air an air cleaner provides to a room or space. When determining CADR, the amount of airflow in a space is taken into account. For example, an air cleaner with a flow rate of 100 cfm (cubic feet per minute) and an efficiency of 50% has a CADR of 50 cfm. Along with CADR, filtration performance is very important when it comes to the air in our indoor environment. Filter performance depends on the size of the particle or fiber, the filter packing density and depth and also the air flow rate.

Constant Air Volume

Constant Air Volume (CAV) systems are a classic solution to meet HVAC needs. Being a simple system in nature, it has become one of the most popular systems overtime. The basic concept of CAV systems is to use a constant supply air volume through the distribution system while heating or cooling the air to meet the spaces needs. There are typically 3 different types of CAV systems; single duct, reheat, and mixed air.

- Single duct systems have one distribution system coming from the source where the air is either heated or cooled. The air, at a constant volume, is then distributed throughout the system to meet the HVAC needs.

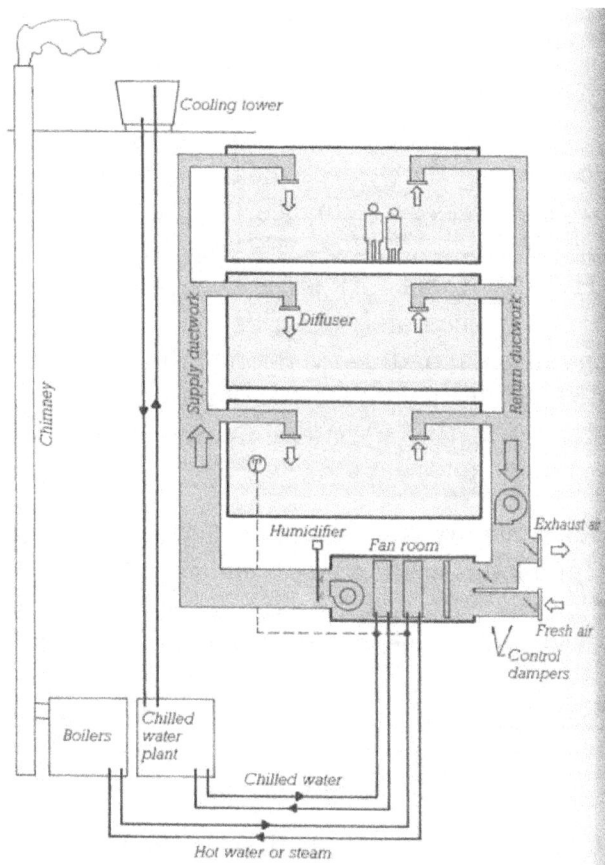

- A reheat system is taking the single distribution system of the single duct method, and adding reheat coils in the ductwork for further temperature control at each individual space. This allows for some spaces to have different supply temperatures than others.

- A mixed air system alternatively has two distribution systems, one for cooling and the other for heating. These two ducts meet at the space and are controlled by a mixing box.

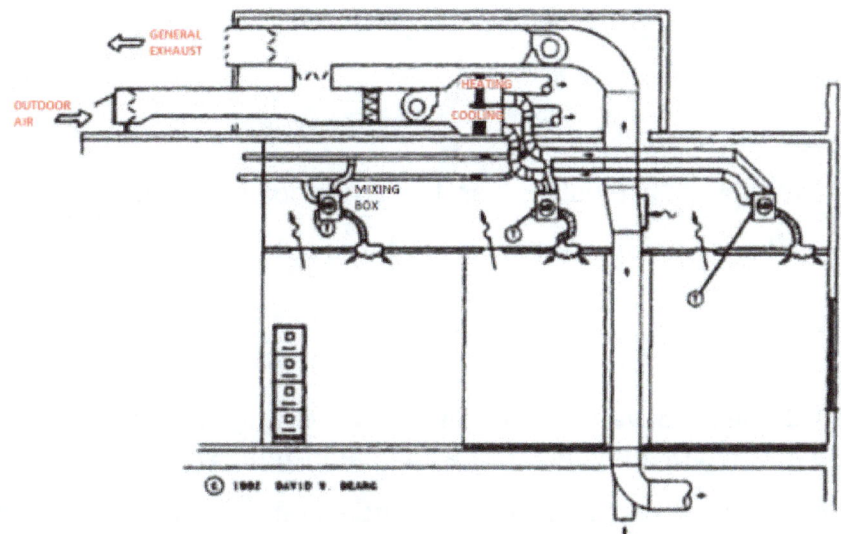

Mixed Air System

This box is a simple valve mechanism that based on the needs of the space, decides how much warm and cool air should be let into the diffuser to accommodate the necessary load.

All of these CAV systems have the capability to control heating, cooling, humidity, gas, and particulates. The pressure and air velocity of the air on the other hand are not really controlled. Since there is a constant volume of air being treated and distributed, there is no chance for pressure changes to be made at individual spaces. This also limits the velocity control capabilities since there is no variation in distribution.

Typical Uses

CAV systems can be used in several different applications depending on whether it is a single duct, reheat, or mixed air system.

Single duct systems are good for areas that require little differences between spaces. For example, a warehouse setting would be good for CAV since it is essential one giant space that has one HVAC load to meet. These systems can also be found in residential areas. Most housing does not have individual space control; rather a central heating system that determines the supply air based on the entire structures needs. Single space buildings, with little variance are the best application for CAV single duct systems.

Reheat systems are good for buildings with a single main space surrounded by smaller user specific spaces. The air-handling unit can control this way the main space and the smaller spaces can be heated or cooled if they have different needs than the main space. An example would be a gymnasium with adjacent locker-rooms and offices.

Mixed Air systems can be appropriate in many different spaces. Since there are mixing boxes at the different spaces, mixed air systems can have individualized control. This makes these systems good for office buildings, schools, and similar structures. In the office building example, individual offices would be able to have control over the temperature in that specific space as opposed to having a single duct system that would only be able to have a single supply condition for all the offices.

Strengths

- Low initial cost

- Easy to design and construct

- Can control ventilation without a separate distribution system and fan equipment

- Easy to expand and modify since ducts can be added onto and mixing boxes can be easily relocated.

Limitations

- Supply air volume is not variable limiting the systems applications (i.e. the system cannot be used in a hospital that is concerned with quarantining spaces).

- Oversized ducts resulting in an increase in initial and operating costs.

- Non-variable supply air volume results in increased fan use since the same amount of air is circulating through the system despite the size of the demand.

- Mixed air systems require two distribution systems increasing cost and spatial requirements.

Parameters

A large part of most air distribution system is the fans. For CAV systems, the ducting and fans significantly contribute to cost and the size of the system. According to ASHRAE STD 90.1, CAV systems under 20,000 cfm are limited to 1.2hp/1000 cfm for fan units. Systems that are over 20,000 cfm are limited to 1.1hp/1,000 cfm for fan units. Below is a graph of typical fan performance curves based on fan horsepower.

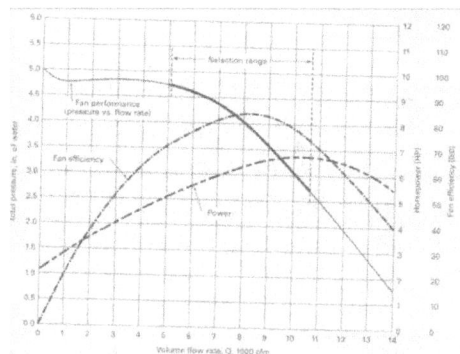

ASHRAE also dictates ductwork sizing based of pressure and velocity classification.
The table below shows several different pressure standards and the possible velocities of the system.

Pressure/velocity classification of ductwork					
Class Type	Class Pressure	Operating Pressure, in. W.C.	Maximum Velocity, ft per min (fpm)	Design Velocity, ft per min (fpm) Main	Branch
1	½" w.c.	−½" to +½"	2000	Up to 2000	Up to 800
2	1" w.c.	−1" to +1"	2500	Up to 2500	Up to 1200
3	2" w.c.	−2" to +2"	2500	Up to 2500	Up to 1500
4	3" w.c.	−3" to +3"	4000	Up to 4000	Up to 2000
5	4" w.c.	−3" to +4"	4000	Up to 4000	Up to 3000
6	5" w.c.	5" to +6"	5000	Up to 5000	Up to 4000
7	6" w.c.	0" to +10"	5000	Up to 5000	Up to 4000

Each class of duct has different sizing and can greatly affect the cost and spatial requirements of the system.

Dedicated Outdoor Air System

A dedicated outdoor air system is an HVAC unit that is installed outside and is often used with other HVAC equipment. DOAS units bring fresh outside air into interior spaces independently from heating or cooling efforts. Addressing ventilation and air conditioning separately can save fan energy while improving indoor air quality.

The type of equipment used with a DOAS may vary depending on brand name and building type. In some instances, a chilled water network is employed to supply the DOAS cooling coils with cold water. Other DOAS models may utilize digital scroll compressors, a microchannel condenser coil, and an electric expansion valve to heat, cool, and dehumidify air.

Benefits

1. While dedicated outdoor air systems could help businesses comply with new commercial HVAC energy efficiency standards, there are a host of other benefits as well:

2. They improve indoor air quality by removing contaminants from outdoor air before channeling it into a building.

3. DOAS units are effective dehumidifiers, and they help prevent moisture-related problems like sick building syndrome and mold growth.

4. Decoupling ventilation from air heating and cooling can also lead to energy savings.

Air Changes Per Hour

An air change is how many times the air enters and exits a room from the HVAC system in one hour. Or, how many times a room would fill up with the air from the supply registers in sixty minutes.

One can then compare the number of room air changes to the Required Air Changes Table below. If it's in the range, you can proceed to design or balance the airflow and have an additional assurance that you're doing the right thing. If it's way out of range, you'd better take another look.

Air Changes Formula

To calculate room air changes, measure the supply airflow into a room, multiply the CFM times 60 minutes per hour. Then divide by the volume of the room in cubic feet:

$$Air\ Changes\ /\ hr = \frac{CFM \times 60\,min}{Volume\ of\ Room}$$

In plain English, we're changing CFM into Cubic Feet per Hour (CFH). Then we calculate the volume of the room by multiplying the room height times the width times the length. Then we simply divide the CFH by the volume of the room.

Here's an example of how a full formula works:

$$Air\ Changes\ /\ hr = \frac{300\ cfm \times 60\ min}{15 \times 20 \times 8}$$

$$Air\ Changes\ /\ hr = \frac{18000}{2400}$$

$$Air\ Changes\ /\ hr = 7.50$$

Now, compare 7.5 air changes per hour to the required air changes for that type of room on the *Air Changes per Hour Table below*. If it's a lunch or break room that requires 7-8 air changes per hour, you're right on target. If it's a bar that needs 15-20 air changes per hour, it's time to reconsider.

Typical Air Changes Per Hour Table

Residential	
Basements	3-4
Bedrooms	5-6
Bathrooms	6-7
Family Living Rooms	6-8
Kitchens	7-8
Laundry	8-9

Light Commercial	
Offices	
Business Offices	6-8
Lunch Break Rooms	7-8
Conference Rooms	8-12
Medical Procedure Offices	9-10
Copy Rooms	10-12
Main Computer Rooms	10-14
Smoking Area	13-15
Restaurants	
Dining Area	8-10
Food Staging	10-12
Kitchens	14-18
Bars	15-20
Public Buildings	
Hallways	6-8
Retail Stores	6-10
Foyers	8-10
Churches	8-12
Restrooms	10-12
Auditoriums	12-14
Smoking Rooms	15-20

Room CFM Formula

Let's look at this engineering formula differently. For example, what if the airflow is unknown and you need to calculate the required CFM for a room? Here is a four-step process on how to calculate the room CFM:

- Step One – Use the above *Air Changes per Hour Table* to identify the required air changes needed for the use of the room. Let's say it's a conference room requiring 10 air changes per hour.

- Step Two - Calculate the volume of the room (L'xW'xH').

- Step Three - Multiply the volume of the room by the required room air changes.

- Step Four - Divide the answer by 60 minutes per Hour to find the required room CFM.

$$Re\,quiredCFM = \frac{VolumeofTheRoomxAirChnagesPerHour}{60\,Minutes}$$

Here's an example of how to work the formula:

$$Re\,quiredCFM = \frac{18x24x10x10xACH}{60\,Minutes}$$

$$Re\,quiredCFM = \frac{43,200CubicFeet}{60\,\min utes}$$

$$Re\,quiredCFM = 720$$

When designing or balancing a system requiring additional airflow for ventilation purposes, remember this room will normally demand constant fan operation when occupied. This may present a problem for other rooms on the same zone, so take that into consideration.

Many of these rooms may require a significant amount of outdoor air. The BTU content of this air has to be included in the heat gain or heat loss of the building when determining the size of the heating and cooling equipment.

Thermal Comfort

When people are dissatisfied with their thermal environment, not only is it a potential health hazard, it also impacts on their ability to function effectively, their satisfaction at work, the likelihood they will remain a customer, and so on.

BS EN ISO 7730 defines thermal comfort as 'that condition of mind which expresses satisfaction with the thermal environment.', i.e. the condition when someone is not feeling either too hot or too cold.

The human thermal environment is not straight forward and cannot be expressed in degrees. Nor can it be satisfactorily defined by acceptable temperature ranges. It is a personal experience dependent on a great number of criteria and can be different from one person to another within the same space.

For example, a person walking up stairs in a cold environment whilst wearing a coat might feel too hot, whilst someone sat still in a shirt in the same environment might feel too cold.

The Health and Safety Executive (HSE) suggest that an environment can be said to achieve 'reasonable comfort' when at least 80% of its occupants are thermally comfortable. This means that thermal comfort can be assessed by surveying occupants to find out whether they are dissatisfied with their thermal environment.

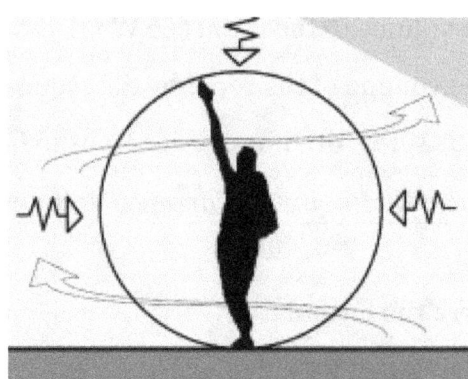

Factors influencing Thermal Comfort

Thermal comfort results from a combination of environmental factors and personal factors.

Environmental Factors

- Air temperature

   The temperature of the air that a person is in contact with, measured by the dry bulb temperature (DBT).

- Air velocity

   The velocity of the air that a person is in contact with (measured in m/s). The faster the air is moving, the greater the exchange of heat between the person and the air (for example, draughts generally make us feel colder).

- Radiant temperature

   The temperature of a persons surroundings (including surfaces, heat generating equipment, the sun and the sky). This is generally expressed as mean radiant temperature (MRT, a weighted average of the temperature of the surfaces surrounding a person, which can be approximated by globe thermometer) and any strong mono-directional radiation such as radiation from the sun.

- Relative humidity (RH)

   The ratio between the actual amount of water vapour in the air and the maximum amount of water vapour that the air can hold at that air temperature, expressed as a percentage. The higher the relative humidity, the more difficult it is to lose heat through the evaporation of sweat.

Personal Factors

- Clothing

   Clothes insulate a person from exchanging heat with the surrounding air and surfaces as well as affecting the loss of heat through the evaporation of sweat. Clothing can be directly controlled by a person (i.e. they can take off or put on a jacket) whereas environmental factors may be beyond their control.

- Metabolic heat

  The heat we produce through physical activity. A stationary person will tend to feel cooler than a person who is exercising.

- Wellbeing and sicknesses

  Such as the common cold or flu which affect our ability to maintain a body temperature of 37° C at the core.

Other contributing factors can include; access to food and drink, acclimatisation (this can be more difficult where there is a high outdoor-indoor temperature gradient) and state of health.

In addition, thermal comfort will be affected by whether a thermal environment is uniform or not. For example, draughts and heaters can create a scorched face/frozen back effect and hot feet/cold head and hands effect.

'Thermal alliesthesia' goes beyond this, proposing that the hedonic qualities of the thermal environment (qualities of pleasantness or unpleasantness, or 'the pleasure principle') are determined as much by the general thermal state of the subject as by the environment itself.

In its simplest form, cold stimuli will be perceived as pleasant by someone who is warm, whilst warm stimuli will be experienced as pleasant by someone who is cold. Introducing a spatial component to this, it can for example be pleasurable to wrap cool hands around a warm mug.

Controlling Thermal Comfort

Thermal comfort can be controlled or adjusted by a number of different measures:

- Environmental monitoring and control (automated or user-controlled systems, active systems such as heating and cooling and passive systems such as shading). NB: User-controlled systems require that users are properly trained.

- Adapting or changing clothing. Businesses can allow people to wear different clothing depending on conditions. They can also provide things like cloak rooms or lockers so that people can change clothes or take off and put down coats. The golden rule is layering, generally 3 layers, and use zips and buttons to regulate temperature.

- Allowing flexible working hours or changing start and finish times.

- Adjusting tasks. For example, allowing breaks or reducing the length of time people are exposed to particular conditions.

- Providing information telling people what sort of conditions to expect so that they can dress and behave appropriately.

- Providing or allowing personal equipment such as desk fans.

- Separating people from sources of discomfort. For example, putting heat generating equipment such as ICT equipment in separate rooms, insulating pipes, preventing draughts and so on. NB: Draughts can be caused by high local surface temperature differences even in a space where there is no air infiltration – for example, a cold down-draught near a window.

- Providing protective clothing (PPE Personal Protective Equipment). This should be a last resort option.

Models

When discussing thermal comfort, there are two main different models that can be used: the static model (PMV/PPD) and the adaptive model.

PMV/PPD Method

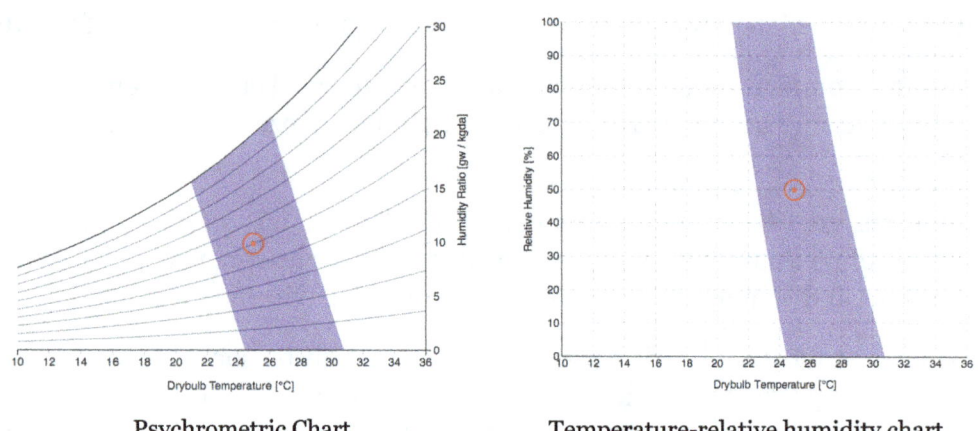

Psychrometric Chart   Temperature-relative humidity chart

Two alternative representations of thermal comfort for the PMV/PPD method

The PMV/PPD model was developed by P.O. Fanger using heat-balance equations and empirical studies about skin temperature to define comfort. Standard thermal comfort surveys ask subjects about their thermal sensation on a seven-point scale from cold (-3) to hot (+3). Fanger's equations are used to calculate the Predicted Mean Vote (PMV) of a large group of subjects for a particular combination of air temperature, mean radiant temperature, relative humidity, air speed, metabolic rate, and clothing insulation. Zero is the ideal value, representing thermal neutrality, and the comfort zone is defined by the combinations of the six parameters for which the PMV is within the recommended limits (-0.5<PMV<+0.5). Although predicting the thermal sensation of a population is an important step in determining what conditions are comfortable, it is more useful to consider whether or not people will be satisfied. Fanger developed another equation to relate the PMV to the Predicted Percentage of Dissatisfied (PPD). This relation was based on studies that surveyed subjects in a chamber where the indoor conditions could be precisely controlled.

This method treats all occupants the same and disregards location and adaptation to the thermal environment. It basically states that the indoor temperature should not change as the seasons do. Rather, there should be one set temperature year-round. This is taking a more passive stand that humans do not have to adapt to different temperatures since it would always be constant.

ASHRAE Standard 55-2010 uses the PMV model to set the requirements for indoor thermal conditions. It requires that at least 80% of the occupants be satisfied.

The CBE Thermal Comfort Tool for ASHRAE 55 allows users to input the six comfort parameters to determine whether a certain combination complies with ASHRAE 55. The results are displayed on a psychrometric or a temperature-relative humidity chart and indicate the ranges of temperature and relative humidity that will be comfortable with the given the values input for the remaining four parameters.

Elevated Air Speed Method

ASHRAE 55 2013 accounts for air speeds above 0.2 metres per second (0.66 ft/s) separately than the baseline model. Because air movement can provide direct cooling to people, particularly if they are not wearing much clothing, higher temperatures can be more comfortable than the PMV model predicts. Air speeds up to 0.8 m/s (2.6 ft/s) are allowed without local control, and 1.2 m/s is possible with local control. This elevated air movement increases the maximum temperature for an office space in the summer to 30° C from 27.5° C (86.0–81.5° F).

Virtual Energy for Thermal Comfort

"Virtual Energy for Thermal Comfort" is the amount of energy that will be required to make a non-air-conditioned building relatively as comfortable as one with air-conditioning. This is based on the assumption that the home will eventually install air-conditioning or heating. Passive design improves thermal comfort in a building, thus reducing demand for heating or cooling. In many developing countries, however, most occupants do not currently heat or cool, due to economic constraints, as well as climate conditions which border lines comfort conditions such as cold winter nights in Johannesburg (South Africa) or warm summer days in San Jose, Costa Rica. At the same time, as incomes rise, there is a strong tendency to introduce cooling and heating systems. If we recognize and reward passive design features that improve thermal comfort today, we diminish the risk of having to install HVAC systems in the future, or we at least ensure that such systems will be smaller and less frequently used. Or in case the heating or cooling system is not installed due to high cost, at least people should not suffer from discomfort indoors. To provide an example, in San Jose, Costa Rica, if a house were being designed with high level of glazing and small opening sizes, the internal temperature would easily rise above 30° C (86° F) and natural ventilation would not be enough to remove the internal heat gains and solar gains. This is why Virtual Energy for Comfort is important.

World Bank's assessment tool the EDGE software (Excellence in Design for Greater Efficiencies) illustrates the potential issues with discomfort in buildings and has created the concept of Virtual Energy for Comfort which provides for a way to present potential thermal discomfort. This approach is used to award for design solutions which improves thermal comfort even in a fully free running building. Despite the inclusion of requirements for overheating in CIBSE, overcooling has not been assessed. However, overcooling can be an issue, mainly in the developing world, for example in cities such as Lima (Peru), Bogota, and Delhi, where cooler indoor temperatures can occur frequently. This may be a new area for research and design guidance for reduction of discomfort.

Standard Effective Temperature

Standard effective temperature (SET*) is a model of human response to the thermal environment. Developed by A.P. Gagge and accepted by ASHRAE in 1986, it is also referred to as the Pierce Two-Node model. Its calculation is similar to PMV because it is a comprehensive comfort index based on heat-balance equations that incorporates the personal factors of clothing and metabolic rate. Its fundamental difference is it takes a two-node method to represent human physiology in measuring skin temperature and skin wettedness.

ASHRAE 55-2010 defines SET as "the temperature of an imaginary environment at 50% relative humidity, <0.1 m/s (0.33 ft/s) average air speed, and mean radiant temperature equal to average air temperature, in which total heat loss from the skin of an imaginary occupant with an activity level of 1.0 met and a clothing level of 0.6 clo is the same as that from a person in the actual environment, with actual clothing and activity level."

Research has tested the model against experimental data and found it tends to overestimate skin temperature and underestimate skin wettedness. Fountain and Huizenga developed a thermal sensation prediction tool that computes SET.

Local Thermal Discomfort

Although thermal comfort is usually discussed for the body as a whole, thermal dissatisfaction may also occur just for a particular part of the body, due to local sources of unwanted heating, cooling or air movement. According to the ASHRAE 55-2010 standard, there are four main causes of thermal discomfort to be considered. A section of the standard specifies the requirements for these factors, that apply to a lightly clothed person engaged in near sedentary physical activity. This is because people with higher metabolic rates and more clothing insulation are less thermally sensitive, and consequently have less risk of thermal discomfort.

Radiant Temperature Asymmetry

Large differences in the thermal radiation of the surfaces surrounding a person may cause local discomfort or reduce acceptance of the thermal conditions. ASHRAE Standard 55 sets limits on the allowable temperature differences between various surfaces. Because people are more sensitive to some asymmetries than others, for example that of a warm ceiling versus that of hot and cold vertical surfaces, the limits depend on which surfaces are involved. The ceiling is not allowed to be more than +5° C (9.0° F) warmer, whereas a wall may be up to +23° C (41° F) warmer than the other surfaces.

Draft

While air movement can be pleasant and provide comfort in some circumstances, it is sometimes unwanted and causes discomfort. This unwanted air movement is called "draft" and is most prevalent when the thermal sensation of the whole body is cool. People are most likely to feel a draft on uncovered body parts such as their head, neck, shoulders, ankles, feet, and legs, but the sensation also depends on the air speed, air temperature, activity, and clothing.

Vertical Air Temperature Difference

Thermal stratification that results in the air temperature at the head level being higher than at the ankle level may cause thermal discomfort. ASHRAE Standard 55 recommends that the difference not be greater than 3° C (5.4° F) for seated occupants or for standing occupants 4° C (7.2° F).

Floor Surface Temperature

Floors that are too warm or too cool may cause discomfort, depending on footwear. ASHRAE 55 recommends that floor temperatures stay in the range of 19–29° C (66–84° F) in spaces where occupants will be wearing lightweight shoes.

Adaptive Comfort Model

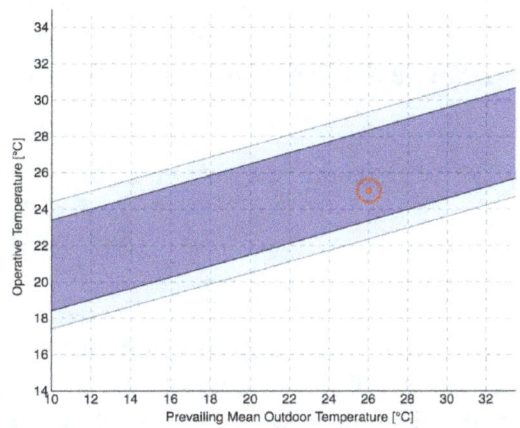

Adaptive chart according to ASHRAE Standard 55-2010

The adaptive model is based on the idea that outdoor climate influences indoor comfort because humans can adapt to different temperatures during different times of the year. The adaptive hypothesis predicts that contextual factors, such as having access to environmental controls, and past thermal history can influence building occupants' thermal expectations and preferences. Numerous researchers have conducted field studies worldwide in which they survey building occupants about their thermal comfort while taking simultaneous environmental measurements. Analyzing a database of results from 160 of these buildings revealed that occupants of naturally ventilated buildings accept and even prefer a wider range of temperatures than their counterparts in sealed, air-conditioned buildings because their preferred temperature depends on outdoor conditions. These results were incorporated in the ASHRAE 55-2004 standard as the adaptive comfort model. The adaptive chart relates indoor comfort temperature to prevailing outdoor temperature and defines zones of 80% and 90% satisfaction.

The ASHRAE-55 2010 Standard introduced the prevailing mean outdoor temperature as the input variable for the adaptive model. It is based on the arithmetic average of the mean daily outdoor temperatures over no fewer than 7 and no more than 30 sequential days prior to the day in question. It can also be calculated by weighting the temperatures with different coefficients, assigning increasing importance to the most recent temperatures. In case this weighting is used, there is no need to respect the upper limit for the subsequent days. In order to apply the adaptive model, there should be no mechanical cooling system for the space, occupants should be engaged in sedentary activities with metabolic rates of 1-1.3 met, and a prevailing mean temperature greater than 10° C (50.0° F) and less than 33.5° C (92.3° F).

This model applies especially to occupant-controlled, natural-conditioned spaces, where the outdoor climate can actually affect the indoor conditions and so the comfort zone. In fact, studies by de Dear and Brager showed that occupants in naturally ventilated buildings were tolerant of a wider range of temperatures. This is due to both behavioral and physiological adjustments, since there are different types of adaptive processes. ASHRAE Standard 55-2010 states that differences in recent thermal experiences, changes in clothing, availability of control options, and shifts in occupant expectations can change people's thermal responses.

Adaptive models of thermal comfort are implemented in other standards, such as European EN 15251 and ISO 7730 standard. While the exact derivation methods and results are slightly different from the ASHRAE 55 adaptive standard, they are substantially the same. A larger difference is in applicability. The ASHRAE adaptive standard only applies to buildings without mechanical cooling installed, while EN15251 can be applied to mixed-mode buildings, provided the system is not running.

There are basically three categories of thermal adaptation, namely: behavioral, physiological, and psychological.

Psychological Adaptation

An individual's comfort level in a given environment may change and adapt over time due to psychological factors. Subjective perception of thermal comfort may be influenced by the memory of previous experiences. Habituation takes place when repeated exposure moderates future expectations, and responses to sensory input. This is an important factor in explaining the difference between field observations and PMV predictions (based on the static model) in naturally ventilated buildings. In these buildings, the relationship with the outdoor temperatures has been twice as strong as predicted.

Psychological adaptation is subtly different in the static and adaptive models. Laboratory tests of the static model can identify and quantify non-heat transfer (psychological) factors that affect reported comfort. The adaptive model is limited to reporting differences (called psychological) between modeled and reported comfort.

Thermal comfort as a "condition of mind" is *defined* in psychological terms. Among the factors that affect the condition of mind (in the laboratory) are a sense of control over the temperature, knowledge of the temperature and the appearance of the (test) environment. A thermal test chamber that appeared residential "felt" warmer than one which looked like the inside of a refrigerator.

Physiological Adaptation

The body has several thermal adjustment mechanisms to survive in drastic temperature environments. In a cold environment the body utilizes vasoconstriction; which reduces blood flow to the skin, skin temperature and heat dissipation. In a warm environment, vasodilation will increase blood flow to the skin, heat transport, and skin temperature and heat dissipation. If there is an imbalance despite the vasomotor adjustments listed above, in a warm environment sweat production will start and provide evaporative cooling. If this is insufficient, hyperthermia will set in, body temperature may reach 40° C (104° F), and heat stroke may occur. In a cold environment, shivering will start, involuntarily forcing the muscles to work and increasing the heat production by up to a factor of 10. If equilibrium is not restored, hypothermia can set in, which can be fatal. Long-term adjustments to extreme temperatures, of a few days to six months, may result in cardiovascular and endocrine adjustments. A hot climate may create increased blood volume, improving the effectiveness of vasodilation, enhanced performance of the sweat mechanism, and the readjustment of thermal preferences. In cold or underheated conditions, vasoconstriction can become permanent, resulting in decreased blood volume and increased body metabolic rate.

Behavioral Adaptation

In naturally ventilated buildings, occupants take numerous actions to keep themselves comfortable when the indoor conditions drift towards discomfort. Operating windows and fans, adjusting blinds/shades, changing clothing, and consuming food and drinks are some of the common adaptive strategies. Among these, adjusting windows is the most common. Those occupants who take these sorts of actions tend to feel cooler at warmer temperatures than those who do not.

These behavioral actions significantly influence energy simulation inputs, and researchers are developing behavior models to improve the accuracy of simulation results. For example, there are many window-opening models that have been developed to date, but there is no consensus over the factors that trigger window opening.

Specificity and Sensitivity

Individual Differences

The thermal sensitivity of an individual is quantified by the descriptor F_s, which takes on higher values for individuals with lower tolerance to non-ideal thermal conditions. This group includes pregnant women, the disabled, as well as individuals whose age is below fourteen or above sixty, which is considered the adult range. Existing literature provides consistent evidence that sensitivity to hot and cold surfaces usually declines with age. There is also some evidence of a gradual reduction in the effectiveness of the body in thermo-regulation after the age of sixty. This is mainly due to a more sluggish response of the counteraction mechanisms in lower parts of the body that are used to maintain the core temperature of the body at ideal values. Seniors prefer warmer temperatures than young adults (76 vs 72 degrees F).

Situational factors include the health, psychological, sociological, and vocational activities of the persons.

Biological Gender Differences

While thermal comfort preferences between sexes seems to be small, there are some differences. Studies have found males report discomfort due to rises in temperature much earlier than females. Males also estimate higher levels of their sensation of discomfort than females. One recent study tested males and females in the same cotton clothing, performing mental jobs while using a dial vote to report their thermal comfort to the changing temperature. Many times, females will prefer higher temperatures. But while females were more sensitive to temperatures, males tend to be more sensitive to relative-humidity levels.

An extensive field study was carried out in naturally ventilated residential buildings in Kota Kinabalu, Sabah, Malaysia. This investigation explored the sexes thermal sensitivity to the indoor environment in non air-conditioned residential buildings. Multiple hierarchical regression for categorical moderator was selected for data analysis; the result showed that females were slightly more sensitive than males to the indoor air temperatures, whereas, under thermal neutrality, it was found that males and females have similar thermal sensation.

Regional Differences

In different areas of the world, thermal comfort needs may vary based on climate. In China the climate has hot humid summers and cold winters, causing a need for efficient thermal comfort. Energy conservation in relation to thermal comfort has become a large issue in China in the last several decades due to rapid economic and population growth. Researchers are now looking into ways to heat and cool buildings in China for lower costs and also with less harm to the environment.

In tropical areas of Brazil, urbanization is causing a phenomenon called urban heat islands (UHI). These are urban areas that have risen over the thermal comfort limits due to a large influx of people and only drop within the comfortable range during the rainy season. Urban heat islands can occur over any urban city or built-up area with the correct conditions. Urban heat islands are caused by urban areas with few trees and vegetation to block solar radiation or carry out evapotranspiration, many structures with a large proportion of roofs, and sidewalks with low reflectivity that absorb heat, high amounts of ground-level carbon dioxide pollution that retains heat released by surfaces, great amounts of heat generated by air-conditioning systems of densely packed buildings, and a large amount of automobile traffic generating heat from engines and exhaust.

In the hot humid region of Saudi Arabia, the issue of thermal comfort has been important in mosques where Muslims go to pray. They are very large open buildings that are used only intermittently (very busy for the noon prayer on Fridays), making it hard to ventilate them properly. The large size requires a large amount of ventilation, but this requires a lot of energy since the buildings are used only for short periods of time. Some mosques have the issue of being too cold from their HVAC systems running for too long, and others remain too hot. The stack effect also comes into play due to their large size and creates a large layer of hot air above the people in the mosque. New designs have placed the ventilation systems lower in the buildings to provide more temperature control at ground level. New monitoring steps are also being taken to improve efficiency.

Thermal Stress

The concept of thermal comfort is closely related to thermal stress. This attempts to predict the impact of solar radiation, air movement, and humidity for military personnel undergoing training exercises or athletes during competitive events. Values are expressed as the wet bulb globe temperature or the discomfort index. Generally, humans do not perform well under thermal stress. People's performances under thermal stress is about 11% lower than their performance at normal thermal wet conditions. Also, human performance in relation to thermal stress varies greatly by the type of task which the individual is completing. Some of the physiological effects of thermal heat stress include increased blood flow to the skin, sweating, and increased ventilation.

Thermal Destratification

One of the fastest growing and most simple energy reduction initiatives which can be installed into both existing and new build facilities is Thermal Destratification, rated by the Carbon Trust as one of the top carbon reducing initiatives for any type of building.

In all buildings the natural process of thermal stratification occurs, which can result in dramatic differences in temperature from floor to ceiling and wall to wall. Thermal stratification is caused by hot air rising up into the ceiling or roof space because it is lighter than the surrounding cooler air. The same applies to cool air falling to the floor as it is heavier than the surrounding warmer air. This means that HVAC systems have to constantly cycle on in order to maintain building interiors at a set and even temperature throughout.

HVAC systems are typically over delivering either heating or cooling to compensate for this stratification phenomenon in an attempt to achieve a required temperature at working/operating level, which is normally only around 1.5 metres to 2 metres from the floor. This costs a lot of money and creates a lot of carbon.

As a result large amounts of wasted heat can build up unseen in ceilings where the difference in temperature can easily rise 14° C or higher than the temperature at floor level depending on floor to ceiling height, and the higher the building the more extreme this temperature differential can be (Building Services Research and Information Association). This heat is also increasing the Delta "T" between inside and outside, accelerating the rate at which hot air escapes through the roof.

This heat can easily be captured and reused by the installation of an efficient destratification fan system, which will balance internal temperatures and thus reduce the operation time and workload required of HVAC systems.

Destratification Technologies

Reducing thermal stratification can be accomplished by controlling the variables that are associated with increased stratification. Since many of the variables, including ceiling height, people and processes, solar gain, and outside weather conditions cannot be controlled, the most common technologies used are related to the building's HVAC (heating, ventilation, and air conditioning) system. One of the cheapest, most effective, and easiest to install technologies are destratification fans, including both axial destratification fans and HVLS (high-volume low-speed) fans.

Axial Destratification Fans

Axial destratification fans are self-contained units that are installed in an array at the ceiling with the goal of blowing conditioned air in the ceiling down to the floor, where people live and work. Because axial fans are designed to blow air straight down at the floor, they can be used in ceiling and roof structures over 100 ft. tall. Because axial destratification fans can achieve destratification with low CFMs, it is imperative that the air leaving the nozzle achieve an air speed at the floor of between 0.2 and 0.5 m/s. The result of this level of air movement is the integration of conditioned air from the ceiling with air at the floor level. Failing to impact the floor will result in destratification of medial layers of air but not achieve destratification at the floor. Since the area around the thermostat will not be destratified in this instance, it is hypothesized that there will be little or no cost savings, as the thermostat will continue to overheat or overcool the room.

An experiment in a room with a 21 ft. ceiling yielded a savings of 23.5% with the use of axial destratification fans.

HVLS Fans

Because of their size, HVLS fans are normally installed in new construction, rather than retrofits, as the roof structure may have to be redesigned to accommodate the increased weight and size. It's not uncommon to require the relocation of lights, due to strobing as large fan blades pass under them, and sprinkler systems, which typically require unobstructed access to the floor to meet fire code. When used in the summer to encourage evaporative cooling, HVLS fans are run forward, blowing air at the floor. When used for destratification in the winter, the fans are run in reverse, blowing air towards ceiling which then circulates around the room. The height at which HVLS fans can be effective is limited compared to axial destratification fans.

Benefits

This method has the most benefits through its application in the heating, ventilation, and air conditioning (HVAC) industry and in heating and cooling for buildings and it has been found that "stratification is the single biggest waste of energy in buildings today."

For Reducing Energy Consumption

By incorporating thermal destratification technology into buildings, energy requirements are reduced as heating systems are no longer over-delivering in order to constantly replace the heat that rises away from the floor area, by redistributing the already heated air from the unoccupied ceiling space back down to floor level, until temperature equalisation is achieved. With regards to cooling destratification systems ensure the cooled air supplied is circulated fully and distributed evenly throughout internal environments, eliminating hot and cold spots and satisfying thermostats for longer periods of time. As a result, destratification technology has great potential for carbon emission reductions due to the reduced energy requirement, and is in turn capable of cutting costs for businesses, sometimes by up to 50%. This is supported by The Carbon Trust which recommends destratification in buildings as one of its top three methods to reduce carbon dioxide emissions.

For Comfort

Destratification naturally increases air movement at the floor, reducing "hot spots" and "cold spots" in a room. It can be used in typically cold areas, like grocery store freezer cases, to warm patrons shopping nearby. In addition, air movement from destratification fans can be used to help meet ASHRAE Standard 62.1 by increasing the amount of air movement at the floor.

Thermal Mass

Thermal mass is a concept in building design that describes how the mass of the building provides inertia against internal temperature fluctuations. This is typically achieved through its ability to absorb unwanted heat during the day and then release it at night with the help of ventilation from cool night air. For a material to provide a useful level of thermal mass, a combination of three basic characteristics is required:

1. A high specific heat capacity; so the heat squeezed into every kg is maximised.

2. A high density; the heavier the material, the more heat it can store by volume.

3. Moderate thermal conductivity – so the rate heat flows in and out of the material is roughly in step with the daily heating and cooling cycle of the building.

Heavyweight construction materials such as masonry and concrete have these characteristics. They combine a high storage capacity with moderate thermal conductivity. This means that heat moves between the material's surface and its interior at a rate that roughly matches the building's daily heating and cooling cycle. Some materials, like wood, have a high heat capacity, but their thermal conductivity is relatively low, limiting the rate at which heat can be absorbed during the day and released at night. Steel can a store a lot of heat, but conducts it too rapidly to be particularly useful, plus comparatively little is used in buildings. However, a modest amount of thermal mass may still be provided if concrete floors are used in steel frame construction, although these are usually limited to a depth of only 100mm and are usually covered by a false ceiling, limiting their ability to absorb and release heat.

Measurement of Thermal Mass

k-values

Part L of the Building Regulations and its associated compliance tools (SAP & SBEM) account for thermal mass using k-values (kJ/m^2K), which provide an indication of the thermal capacity per square metre of floor or wall. Lightweight walls have a low k-value of around 10 kJ/m^2K, whilst for heavyweight walls it can be up to 230 kJ/m^2K.

Admittance Values

Describing a material or construction as having high, medium or low thermal mass gives a useful indication of its ability to store heat, as does its k-values. But, in order to get a better idea of how effective it is likely to be in practice, there are a couple of other important factors that need to be considered. These are firstly the length of time available to get heat in and out of the material, which is typically assumed to be 24 hours, and secondly, the rate of heat flow at its surface i.e. through carpet, plasterboard, tiles etc. These factors are accounted for in admittance values, which provide a simple means of assessing the approximate in-use thermal mass performance of walls and floors, making it a more sophisticated metric than k-values etc.

Decrement

Admittance values and k-values relate to the absorption of heat inside buildings, which is the most important use of thermal mass. There is however another thermal mass related property called decrement, which can influence summertime performance to a limited extent. Decrement describes the way in which the density, heat capacity and thermal conductivity of an external wall (for example), can slow the passage of heat from the sun as it passes from the outside to the inner surface of the wall (decrement delay), and also reduce those gains as they pass through it (decrement factor).

Working of Thermal Mass

Thermal mass acts as a thermal battery. During summer it absorbs heat during the day and releases it by night to cooling breezes or clear night skies, keeping the house comfortable. In winter the same thermal mass can store the heat from the sun or heaters to release it at night, helping the home stay warm.

Thermal mass is not a substitute for insulation. Thermal mass stores and re-releases heat; insulation stops heat flowing into or out of the building. A high thermal mass material is not generally a good thermal insulator.

Thermal mass is particularly beneficial where there is a big difference between day and night outdoor temperatures.

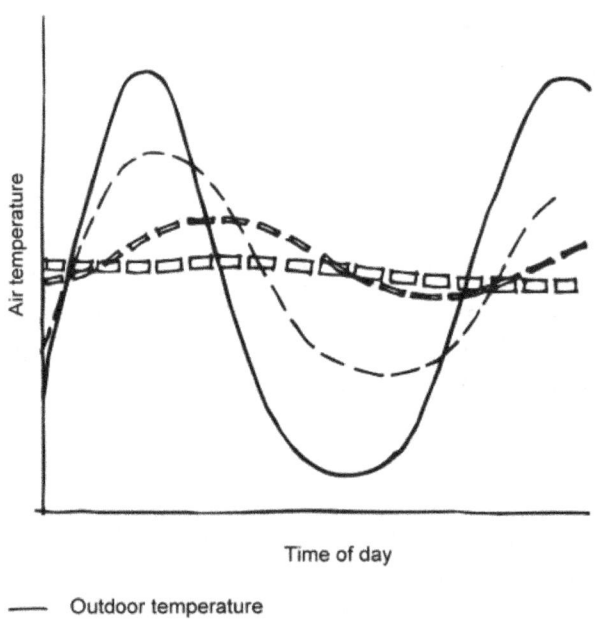

Time of day

——— Outdoor temperature

— — Light timber-framed building

━━ Heavy building with external insulation

▢▢ Heavy building set into and partially covered with earth

Daily temperature fluctuations for different construction methods

Correct use of thermal mass can delay heat flow through the building envelope by as much as 10–12 hours, producing a warmer house at night in winter and a cooler house during the day in summer.

A high mass building needs to gain or lose a large amount of energy to change its internal temperature, whereas a lightweight building requires only a small energy gain or loss to change the air temperature. This is an important factor to consider when choosing construction systems and assessing climate change adaptation.

Winter

Allow thermal mass to absorb heat during the day from direct sunlight or from radiant heaters. It re-radiates this warmth back into the home throughout the night.

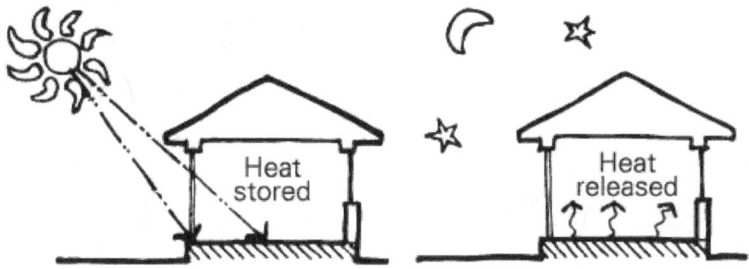

Summer

Allow cool night breezes and convection currents to pass over the thermal mass, drawing out all the stored energy. During the day protect the thermal mass from excess summer sun with shading and insulation if required.

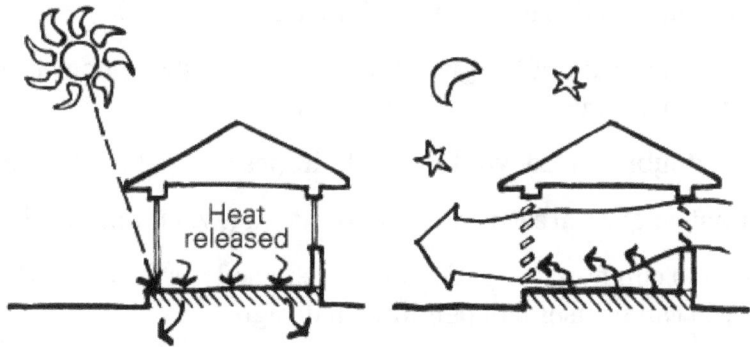

Using Thermal Mass Effectively

Thermal mass is most appropriate in climates with a large diurnal temperature range. As a rule of thumb, diurnal ranges of less than 6° C are insufficient; 7°–10° C can be useful depending on climate; where they exceed 10° C, high thermal mass construction is desirable. Exceptions to the rule occur in more extreme climates.

In cool or cold climates where supplementary heating is often used, houses benefit from high mass construction regardless of diurnal range (e.g. Hobart 8.5° C). In tropical climates with diurnal ranges of 7°–8° C (e.g. Cairns 8.2° C) high mass construction can cause thermal discomfort unless carefully designed, well shaded and insulated.

Always use thermal mass in conjunction with sound climate-appropriate passive design.

Glass to Mass Ratios for Different Climates

Glass to mass ratios compare the area of solar exposed, passively shaded, north-facing glazing to the area of exposed, insulated internal mass (walls and floor), to avoid overheating passive solar houses. The graph below shows recommended glass to mass ratios for Australian capital cities.

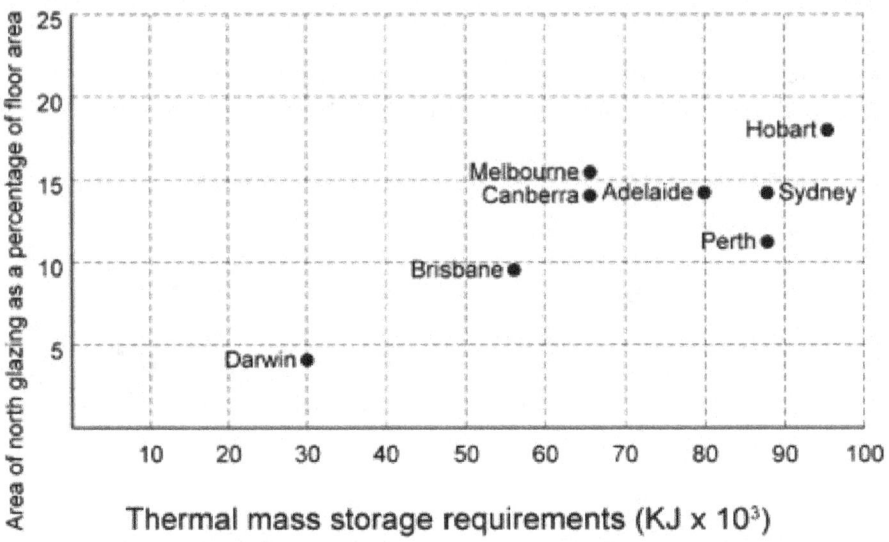

Glass to Mass Ratios in Australian Cities

Rule-of-thumb Glass to Mass Ratios for different Climates

- Cold and alpine climates: double glazed areas of 20–25% of floor area (snug drapes and pelmets should also be used).

- Cool temperate: double glazed windows with drapes and pelmets 15–20% of floor area.

- Temperate climates: glass area 12–15% of floor area (17% if double glazed).

- Cooling dominated climates: the ratio of solar exposed north-facing glass should be at least 6% but up to 10% can be useful depending on design.

- Cooling only climates: solar exposed glass should be avoided; low mass construction with high level ventilation is usually best. Earth coupled slabs can add useful 'heat wicking' properties to thermal mass where mass covered ground temperatures at 1.5m depth remain below 19° C in summer, i.e. not Darwin.

These ratios should be varied according to:

- Solar availability (access and incidence)

- Diurnal temperature ranges

- Type and orientation of glazing and shading (ambient and diffuse gains).

These rules apply only to predominantly north glazed passive designs with guaranteed solar access. Modelling with energy rating software is the only reliable way to validate them.

Typical Applications

In rooms with good access to winter sun it is useful to connect the thermal mass to the earth. The most common example is slab-on-ground construction. Less common examples are brick or earthen floors, earth-covered housing or green roofs.

A slab-on-ground is preferable to a suspended slab in most climates because it has greater thermal mass due to direct contact with the ground. This is known as earth coupling. Deeper, more stable ground temperatures rise beneath the house because its insulating properties prevent heat loss. The slab assumes this higher temperature which can range from 16° to 19° C.

In summer, the earth has the capacity to 'wick' away substantial heat loads. It also provides a cool surface for occupants to radiate heat to (or conduct to, with bare feet). This increases both psychological and physiological comfort.

In winter, the slab maintains thermal comfort at a much higher temperature with no heat input. The addition of passive solar or mechanical heating is then more effective due to the lower temperature increase required to achieve comfortable temperatures.

Use surfaces such as quarry tiles or simply polish the concrete slab. Do not cover areas of the slab exposed to winter sun with carpet, cork, wood or other insulating materials: use rugs instead.

The vertical edges of a slab-on-ground are required to be insulated in Zone 8 (cold climate) or when in-slab heating or cooling is installed within the slab.

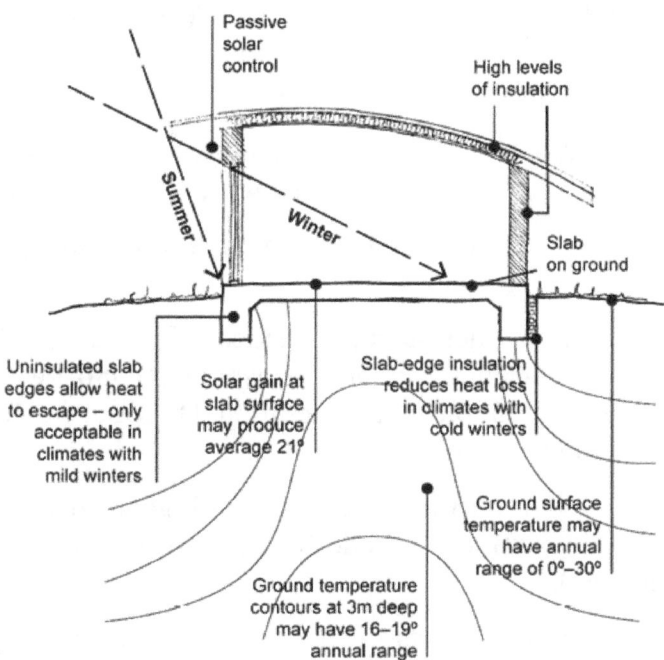

Insulate slab edges in cold climates or where in-slab heating or cooling is installed within the slab.

The whole slab must be insulated from earth contact in cold climates and regions with low earth temperatures at 3m depth (e.g. Tasmania) or hot, humid climates with high earth temperatures (e.g. Darwin).

Consider termite proofing when designing slab edge insulation. Take care to ensure that the type of termite management system selected is compatible with the slab edge insulation.

Masonry walls also provide good thermal mass. Recycled materials can be used (e.g. reused bricks). Avoid finishing masonry walls with plasterboard as this insulates the thermal mass from the interior and significantly reduces its capacity to absorb and re-release heat.

Reverse brick veneer is an example of good thermal mass practice for external walls because the mass is on the inside and externally insulated. In traditional brick veneer, the mass of the brick makes no contribution to thermal storage because it is insulated from the inside and not the outside.

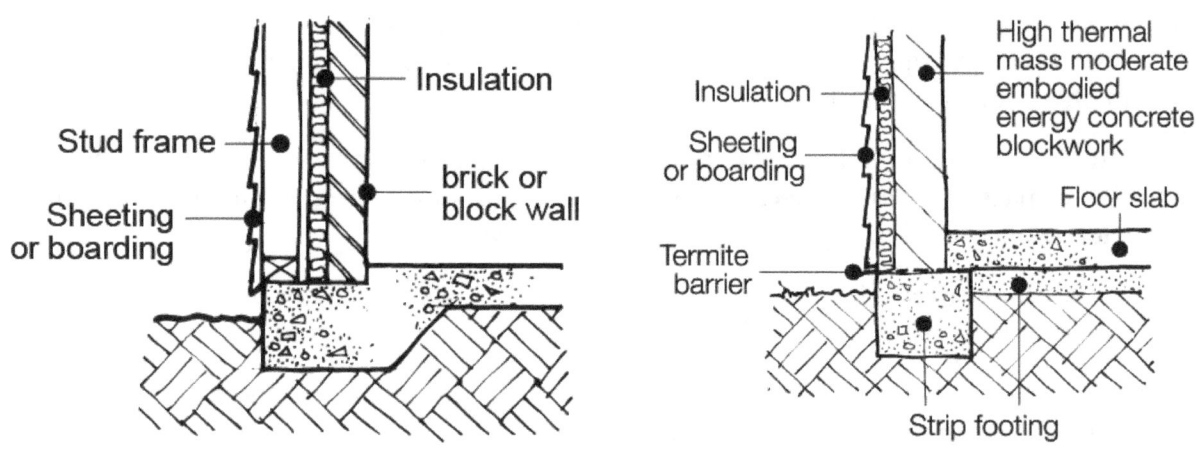

Thermal mass floor and lightweight frame   Thermal mass walls and floor slab

Locating Thermal Mass

To determine the best location for thermal mass you need to know if your greatest energy consumption is the result of summer cooling or winter heating.

- Heating: Locate thermal mass in areas that receive direct sunlight or radiant heat from heaters.

- Heating and cooling: Locate thermal mass inside the building on the ground floor for ideal summer and winter efficiency. The floor is usually the most economical place to locate heavy materials, and earth coupling gives additional thermal stabilization in both summer and winter in these climates.

  o Locate thermal mass in north-facing rooms with good solar access, exposure to cooling night breezes in summer, and additional sources of heating or cooling (heaters or evaporative coolers).

  o Locate additional thermal mass near the centre of the building, particularly if a heater or cooler is positioned there. Feature brick walls, slabs, water features and large earth or water-filled pots can be used.

- Cooling: Protect thermal mass from summer sun with shading and insulation if required. Allow cool night breezes and air currents to pass over the thermal mass, drawing out all the stored energy.

Avoiding Thermal Mass

Avoid use in rooms and buildings with poor insulation from external temperature extremes and rooms with minimal exposure to winter sun or cooling summer breezes.

Thermal mass can increase energy use when used in rooms where auxiliary heating or cooling is the only means of adjusting the temperature because it slows the response times.

Careful design is required if locating thermal mass on the upper levels of multi-storey housing in all but cold climates, especially if these are bedroom areas.

Natural convection creates higher upstairs room temperatures and upper level thermal mass absorbs this energy. On hot nights upper level thermal mass can be slow to cool, causing discomfort. The reverse is true in winter.

Specific Climate Responses

Climatic consideration is critical in the effective use of thermal mass. It is possible to design a high thermal mass building for almost any climate but the more extreme climates require very careful design.

Will the current use of thermal mass still be appropriate in 20 or 30 years' time?

Think about the impact of predicted changes in climate due to global warming. Will the current use of thermal mass still be appropriate in 20 or 30 years' time if temperatures rise and diurnal ranges are reduced? This is a particularly important issue in tropical climates where temperatures are already close to the upper comfort level. For the main features of these climates see Design for climate.

Hot Humid Climates

Use of high mass construction is generally not recommended in hot humid climates due to their limited diurnal range. Passive cooling in this climate is usually more effective in low mass buildings.

Thermal comfort during sleeping hours is a primary design consideration in tropical climates. Lightweight construction responds quickly to cooling breezes. High mass can completely negate these benefits by slowly re-releasing heat absorbed during the day.

Warm Humid and Warm/Mild Temperate Climates

Maintaining thermal comfort in these benign climates is relatively easy. Well-designed houses should require little if any supplementary heating or cooling. In fact, 7–8 Nationwide House Energy Rating Scheme (NatHERS) stars can be achieved at relatively low cost.

The predominant requirement for cooling in these climates is often suited to lightweight, low mass construction. High mass construction is also appropriate but requires sound passive design to avoid overheating in summer.

In multi-level design, high mass construction should ideally be used on lower levels to stabilise temperatures. Low mass on the upper levels ensures that as hot air rises (through convective ventilation) it is not stored in the upper level as it leaves the building.

This is particularly important if sleeping spaces are located on upper levels. Ground and first floor spaces should be capable of zoning (closing off) to prevent temperature stratification in winter.

Cool Temperate and Alpine Climates

Winter heating is the main need in these climates although some summer cooling is generally required. Ceiling fans usually provide adequate cooling in these low humidity climates.

High mass construction combined with sound passive solar design and high level insulation is an ideal solution. Good solar access is required in winter to heat the thermal mass. Glass to mass ratios are critical.

Insulate slab edges and the underside of suspended slabs in colder climates. It is advisable to insulate the underside of a slab-on-ground in extremely cold climates.

Buildings that receive little or no passive solar gains can still benefit from high mass construction if they are well insulated. However, they respond slowly to heating input and are best suited to homes with high occupation rates.

Auxiliary heating of thermal mass is ideally achieved with efficient or renewable energy sources such as solar, gas or geothermal powered hydronic systems. In-slab electric resistance systems are slow responding and cause higher greenhouse gas emissions.

Use a solar conservatory in association with thermal mass to increase heat gains. A solar conservatory is a glazed north-facing room that can be closed off from the dwelling at night. Shade the conservatory in summer and provide high level ventilation to minimise overheating. Reflective internal blinds also reduce winter heat loss.

Hot Dry Climates

Both winter heating and summer cooling are very important in these climates. High mass construction combined with sound passive heating and cooling principles is the most effective and economical means of maintaining thermal comfort.

Diurnal ranges are generally quite significant and can be extreme. High mass construction with high insulation levels is ideal in these conditions.

Where supplementary heating or cooling is required, locate thermal mass where it is exposed to radiation from heaters or cool air streams from evaporative coolers. The mass moderates temperature variations between high/low or on/off and lowers the level and duration of auxiliary requirements while increasing thermal comfort. With the low humidity in these climates, ceiling fans generally provide adequate cooling comfort in a well-designed home.

Underground or earth covered homes give protection from solar radiation and provide additional thermal mass through earth coupling to stabilize internal air temperatures.

Renovations and Additions

When renovating, remove carpet or insulating coverings from concrete slabs that are exposed to winter sun. The slab surface can be tiled or cut and polished to give an attractive and practical finish. Thermal mass can also be increased by adding brick or stone veneers to existing interior walls.

In some cases it may be necessary to reduce the amount of thermal mass exposed to the building interior where insufficient passive heating or cooling is available to maintain comfort. In such cases, additional auxiliary heating or cooling is required. To isolate existing mass, line the interior wall surface with sheet insulation materials and plasterboard.

If planning an addition, engage a thermal performance assessor to model your whole home to identify strengths and weaknesses in relation to windows (orientation and size) and appropriate levels of thermal mass. This model identifies problem areas that might be able to be overcome by adding (or deleting) new rooms.

Heating Dominated Climates

For heating dominated climates, add thermal mass where winter solar access is already available, such as those buildings with good northerly access. This may be achieved by exposing existing concrete as above or adding thermal mass to walls.

Where the existing floor is slab-on-ground, non-loadbearing walls can be built directly on the concrete slab, after engineering checks are carried out. If the existing building has a raised timber floor it is often practical to combine reverse brick veneer with a retrofitted suspended concrete slab. The underside should be insulated and well ventilated if not earth coupled.

It may be necessary to consider revising the layout of the house, 'turning the house around' to place living areas to the north.

Thermal mass should be located near a heater.

Cooling Dominated Climates

For cooling dominated climates, thermal mass must be protected from summer sun and exposed to cooling night breezes.

Add shading to protect thermal mass from summer sun both internally and externally, particularly outside windows and in uninsulated double brick walls. Thermal mass's ability to absorb and re-radiate heat over many hours means that in summer or hot climates it can be a source of unwelcome heat long after the sun has set.

Other Thermal Mass Options

Introduce thermal mass within lightweight structures by using isolated masonry walls, water filled containers, phase change materials (PCMs) or lightweight steel-framed concrete floors.

Internal or enclosed water features such as pools can also provide thermal mass but require good ventilation and must be capable of being isolated, as evaporation can absorb heat in winter and create condensation problems year round.

Air enters this building across the pool (thermal mass) through a semi-enclosed courtyard. It is evaporatively cooled before entering the building. This 'coolth' can be stored in thermal mass.

Roof-mounted solar heating of pools is relatively inexpensive and can be used in conjunction with hydronic heating systems or water storage containers to heat thermal mass in winter or (in reverse) supply radiant cooling to night skies in summer. This method can resolve situations where direct solar access for passive heating is unachievable or where conventional thermal mass is inappropriate (e.g. pole homes).

Thermal Mass Properties

The following characteristics determine thermal mass performance:

- High density: The more dense the material (i.e. the less trapped air), the higher its thermal mass. For example, concrete has high thermal mass, aerated concrete (AAC) blocks have moderate to low thermal mass, and insulation has almost none.

- Good thermal conductivity: To be effective in most climates, thermal mass should have the capacity to absorb and re-emit close to its full heat storage capacity in a single diurnal cycle. If conductivity is too low, passive heating can escape from your home before being absorbed. If conductivity is too high (e.g. steel), stored heat is re-released before it is most needed in the colder part of the night. The same applies to passive cooling only in day–night reverse.

- For example, rubber has high density but is a poor conductor of heat. Brick and concrete have high density and are reasonably good conductors.

- Appropriate thermal lag: The rate at which heat is absorbed and re-released by uninsulated material is referred to as thermal lag. Lag is dependent on conductivity, thickness, insulation levels and temperature differences either side of the wall. Consideration of lag times is important when designing thermal mass, especially with thick uninsulated external wall systems like rammed earth, mud brick or rock.

  In moderate climates, a 24 hour lag cycle is ideal. In colder climates subject to long cloudy periods, lags of up to seven days can be useful, providing there is additional solar exposed glazing to 'charge it' in sunny weather. The table indicates lag times for common materials.

Time lag figures for various materials	
Material thickness (mm)	**Time lag (hours)**
Double brick (220)	6.2
Concrete (250)	6.9
Autoclaved aerated concrete (200)	7.0
Mud brick/adobe (250)	9.2
Rammed earth (250)	10.3
Compressed earth blocks (250)	10.5
Sandy loam (1000)	30 days

Thermal lag influences internal–external heat flow through walls.

Thermal lag influences internal–external heat flow through walls. Rammed earth, rock and mud brick have a low insulation value and rely on thicknesses of 300mm or more to increase thermal lag. While this is often adequate in mild climates, these systems require external insulation in cool and cold climates where lag times are reduced by increased internal–external temperature differences (known as delta T).

- Low reflectivity: Dark, matt or textured surfaces absorb and re-radiate more energy than light, smooth, reflective surfaces (if there is considerable thermal mass in the walls, a more reflective floor will distribute heat to the walls).

- High volumetric heat capacity (VHC): The table below compares the thermal mass performance (or VHC) of some common materials. The amount of useful thermal storage is calculated by multiplying the VHC by the total accessible volume of the material, i.e. the volume of material that has its surface exposed to a source of heating or cooling.

Water has the highest VHC of any common material. The table tells us that it takes 4186KJ of energy to raise the temperature of one cubic metre of water by one degree C, whereas it takes only 2060KJ to raise the temperature of an equal volume of concrete by the same amount. In other words, water has around twice the heat storage capacity of concrete. The VHC of rock usually ranges between brick and concrete depending on density.

The VHC of any material is reduced or even eliminated if the material is covered with linings such as carpets, plasterboard, timber.

Thermal mass for various materials	
Material	**Thermal mass (volumetric heat capacity, KJ/m^3.k)**
Water	4186
Concrete	2060
Sandstone	1800
Compressed earth blocks	1740
Rammed earth	1673
Fibre cement sheet (compressed)	1530

Thermal mass for various materials	
Material	**Thermal mass (volumetric heat capacity, KJ/m³.k)**
Brick	1360
Earth wall (adobe)	1300
Autoclaved aerated concrete	550

Some thermal mass materials, such as concrete and brick, have high embodied energy when used in the quantities required. Consider the lifetime energy impact of thermal mass materials: will the savings in heating and cooling energy be greater than the embodied energy content over the life of the building? Can lower embodied materials such as water or recycled brick be used?

In addition, poor design of thermal mass may result in increased heating and cooling energy use on top of the embodied energy content.

Phase Change Materials

There is growing interest in the use of phase change materials (PCMs) as a lightweight thermal mass substitute in construction. All materials require a large energy input to change state (i.e. from a solid to a liquid or a liquid to a gas). This energy does not change their temperature — only their state. For that reason, it is called 'latent' (i.e. latent heat of melting or vaporisation). Phase change temperatures vary enormously between materials.

Phase change materials, or PCMs, may be a useful lightweight substitute for thermal mass.

Materials that melt between 25° and 35° C are very useful for storing passive solar gains. Any temperature increase over a desired thermal comfort level is absorbed by the PCM as it melts. This energy stays stored until the PCM starts to solidify again as temperatures drop at night. As it solidifies, it releases the stored heat.

Commonly used PCMs include paraffin wax and a variety of benign salts. Many are available in Australia. PCMs are currently expensive compared to conventional thermal mass but can reduce costs through space and structural savings. They are an ideal way to install mass in existing buildings and are particularly useful in lightweight buildings where cost savings are often achieved.

The PCM market is developing rapidly so current suppliers are best found through an internet search. Some PCMs crystallise after many cycles of phase change, which renders them useless. Get a guarantee from your supplier that their product does not do this.

At least one company manufactures building products that integrate phase change microcapsules into their structure, including plasterboard and AAC blocks although this product is currently (in 2012) prohibitively expensive. Gypsum plaster, paints and floor screeds have the potential to contain PCMs and many such applications are likely to appear on the market over the next few years as the technology offers the prospect of lightweight buildings that can behave with characteristics associated with 'traditional' thermal mass. For example, the thermal capacity of a 13mm thick plaster layer with 30% microcapsule content is claimed to be equivalent to that of a 150mm thick masonry wall.

Use of PCMs can be very helpful on severely constrained sites where thermal mass would otherwise be difficult to install.

PCMs or Water as Flexible Mass Options for Climate Change

Because the role of thermal mass is primarily one of heat storage in heating climates, it is likely to become less useful as the climate warms within the lifespan of the home. Additionally, it could become a cooling liability as the prevalence of favourable night-time cooling conditions diminishes.

Replacement of conventional masonry with PCMs could present a solution to the challenge of designing for current and future climates. PCM thermal mass could easily be removed from the building, initially perhaps on a seasonal basis as the climate changes and ultimately become permanent should peak predicted levels of warming occur.

Low Cost Mass Options for Upper Storeys

PCMs or water filled containers have much greater thermal storage capacity than masonry and can be used as a mass substitute. PCMs are much lighter than masonry. Water has double the storage capacity of concrete and because of convection within the container, penetration rates are substantially higher. Thus water can supply similar storage capacity to masonry with significantly less mass and bulk. Accordingly, both can be cost effective mass options for upper storeys because they require no (or less) additional structural support.

Water filled balustrades provide abundant thermal mass as part of this mezzanine balcony

Mobile Thermal Mass

An added benefit arising from water or PCM substitutes for masonry in upper levels is their potential for mobility. Water containers could be drained and PCM containers moved outside should they become a thermal comfort liability in any season or lifestyle pattern.

Mobile PCMs and water can be placed in ideal solar gain positions by day and moved to convenient heating locations at night. Similarly, they can be placed in breeze paths or outside night sky radiation positions to cool at night and moved to warmer rooms during the day to even out diurnal fluctuations.

Multi-storey Buildings

Multi-storey buildings usually include dense concrete cores, particularly for elements like stair and lift wells. Multi-unit dwellings also demand good fire separation that is often most economically and effectively provided by using concrete construction, whether precast, in situ or as blockwork. In each case the high density concrete elements make excellent thermal mass. Its situation in the core of an apartment or as party walls in well-insulated units is good placement for thermal mass and should be incorporated as such into the overall design strategy.

Thermal Mass Checklist

The simple rules of thumb set out here help determine appropriate thermal mass levels in different climate zones — heating dominated temperate and cold climates, cooling dominated temperate climates, and heating dominated climates with no northerly solar access. Mass levels vary according to:

- Solar access (glazing type, orientation, area and shading)

- Cool breeze and cool night air access (including mechanical)

- Diffuse and ambient heat gains in summer

- Night-time sleeping comfort

- Occupation patterns and heating/cooling system use

- Seasonal extremes (climate zone).

The average diurnal range is a useful indicator of appropriate thermal mass levels in a house:

- Low mass construction generally performs best where diurnal ranges are consistently 6° C or less (coastal, temperate climates).

- Moderate mass is best for a 6°–10° C diurnal range (slab-on-ground, lightweight walls such as brick veneer).

- High mass construction is desirable for a diurnal range over 10° C (slab-on-ground and some or all high mass walls).

However, simulation modelling with house energy rating software is the only way to validate these guidelines for a specific house design and climate zone.

Heating Dominated Temperate and Cold Climates

- In the quantities present in most standard construction (e.g. brick veneer with exposed, uncarpeted concrete slab-on-ground), thermal mass is useful for evening out diurnal temperature ranges.

- Greater quantities of both thermal mass and passive heating and cooling are required to moderate temperature cycles up to one week (e.g. slab-on-ground with masonry walls or earth bermed).

- Very high levels of thermal mass (e.g. earth covered buildings) can even out summer–winter ranges if well-designed.

Earth bermed south wall

Cooling Dominated Temperate Climates

Where cooling loads are equal to or greater than heating loads, low to moderate levels of mass are often preferable.

- Earth coupled slabs can moderate diurnal cycles by absorbing summer heat loads, providing a radiant cooling source, and storing winter solar gains for limited periods.

- Non earth coupled mass can overheat during summer days, leaving an undesirable radiant heat source at night — particularly in upper level bedrooms.

- Well-designed or located thermal mass walls backing onto conditioned spaces in hybrid designs (i.e. using both passive and active cooling) can create a source of radiant cooling (you radiate to the mass). This improves sleeping comfort and allows cooling to be turned off or down. It is ideal when combined with ceiling fans in open ventilated south-facing or ground floor sleeping areas.

Heating Dominated Climates with no Northerly Solar Access

In heating dominated climates where northerly solar access is unavailable, westerly sun (if available) is desirable providing:

- Glazing is adequately actively shaded in summer

- Double glazing and drapes with pelmets are used to compensate for reduced heat gain or higher heat loss (3 heat gain hours versus 21 heat loss hours)

- Thermal mass is reduced where limited solar heat gains necessitate additional heating.

Actively shaded, double glazed west-facing windows with solar access are particularly useful for meeting variable heating and cooling needs in spring and autumn. Easterly sun can be less effective for heating in some cooler climates due to morning fog.

Stack Effect

The stack effect is a buoyancy-driven phenomenon that commonly occurs in high-rise buildings. This physical phenomenon typically arises in regions experiencing extreme climatic conditions. The main driver behind the stack effect phenomenon is the temperature difference between the interior of the building and the external environment. Also, the impact of wind pressure acting on the external envelope of the building should not be ignored as it can be a significant contributor to the overall building performance.

Buoyancy-driven Processes

In cold regions, the relatively warmer indoor air of a tall building rises due to buoyancy forces, creating a pressure difference that tries to draw air in at the bottom of the building and pushes air out at the top levels. The cold air that has been drawn in is then heated up by the building services, closing the cycle of the classical stack effect process.

On the other hand, in hot climates, a reverse process can be observed: in this case the relatively cooler indoor air of a high-rise building precipitates, creating a pressure build-up around the lower portion of the building that pushes the air out and draws the air in at the upper levels. Again, the warm air that has been drawn in is then cooled down by the building services, closing the cycle of the 'reverse' stack effect process.

It should be noted that in both types of stack effect processes there is usually a change in sign (direction) of the pressure gradient at the building envelope: this indicates that, at some point along the height of the building, there is a zone of neutral pressure where the internal and external pressures are perfectly equalized.

Wind-driven Processes

Additionally to the buoyancy force, it is also important to take into account the impact of the wind pressure acting on the envelope of the building. It is well known that, due to the shape of the atmospheric boundary layer, the lower and upper levels of tall buildings are subjected to rather different wind speeds (wind pressures). This can have a non-negligible impact on infiltration and exfiltration through the skin of the building; in particular, positive external wind pressures have the ability of enhancing infiltration and counteracting exfiltration whilst negative external wind pressures have the ability of enhancing exfiltration and counteracting infiltration.

Design Challenges

The main design challenges associated with the stack effect phenomenon in tall buildings are:

- Elevator doors operation: elevator doors, due to excessive pressure difference across them, could malfunction and not operate correctly within their guide rails;

- Swing doors operation: users, due to the excessive pressure difference across swing doors, could experience difficulties in opening/closing them;

- Uncomfortable and excessive air flow movement: this has the potential to occur within key occupied spaces such as lobbies, corridors, atria, etc.;

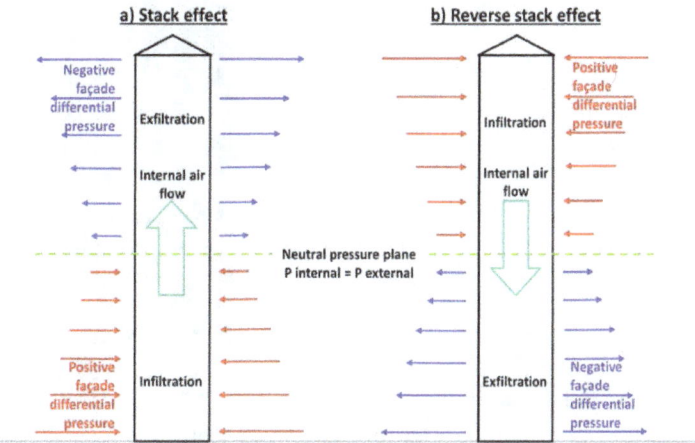

Principle stack effect diagrams

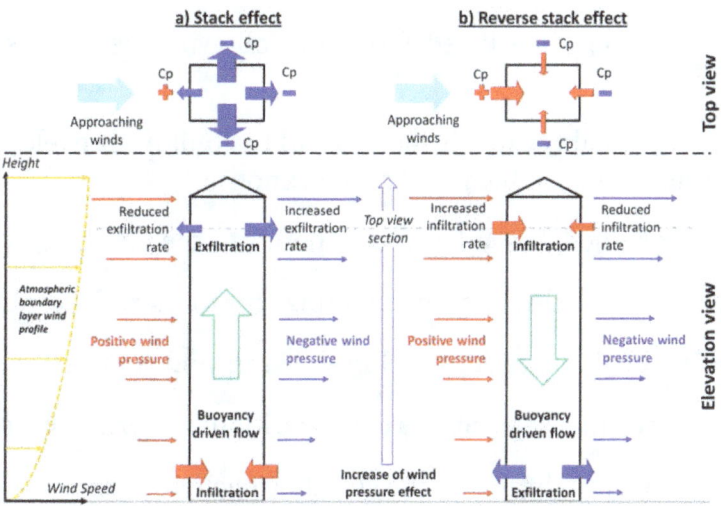

Wind pressure impact over building stack effect

- Propagation spreading of smoke, odors and other unwanted contaminants throughout the building;

- Inefficient heating/cooling strategy: because of the excessive infiltration of cold (hot) ambient air into the lower (upper) levels of the tall buildings, extra energy supply is likely to be required to heat up (cool down) such spaces; Flow-borne noise: high speed air flow through narrow gaps (e.g., door gaps, louvers of the shafts or natural ventilation openings) could be the cause of narrowband high pitch whistling which in turn can create discomfort to the occupants of the tall building;

- Fire strategy: an excessive air flow movement within the tall building could increase the propagation rate of smoke and fire. Also, excessive deviation from prescribed pressurization levels along the main evacuation paths (e.g., stairwells and corridors) due to stack effect, could impede smooth occupants' evacuation procedures to take place.

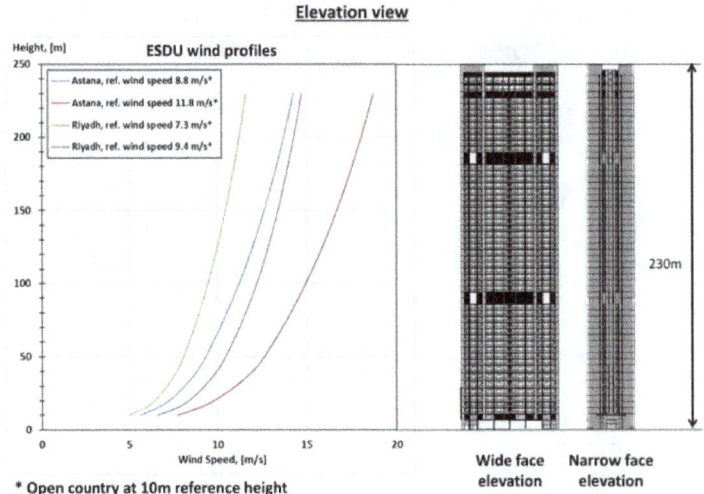

Building elevation and wind profiles

Preventive Measures

In order to prevent some of the issues listed above, the following measures should be considered/implemented:

- Improvement of the quality and air tightness of the building envelope (from design to on-site QA/QC through detailed technical specification);

- Installation of revolving doors at key access points to the building;

- Implementation of vestibules between building entrances and elevator banks;

- Introduction of vertical separations within elevators and stairwell shafts;

- Introduction of horizontal separations (e.g., additional internal partitioning);

- Improvement of the air tightness of the elevator machine room;

- Implementation of mild pressurization within elevators and stairwell shafts;

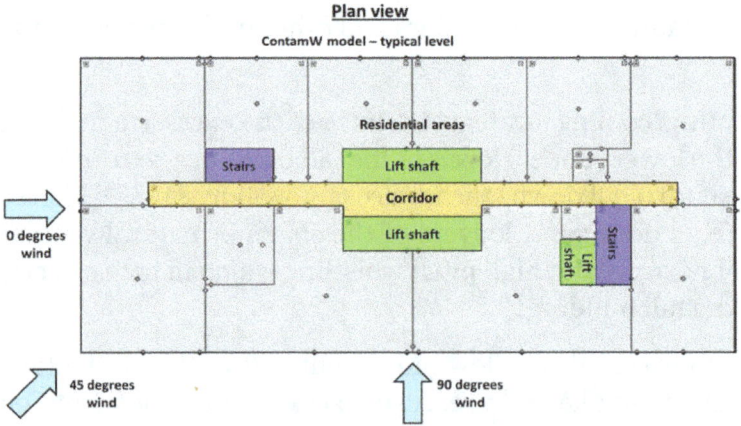

Building plan and approaching wind

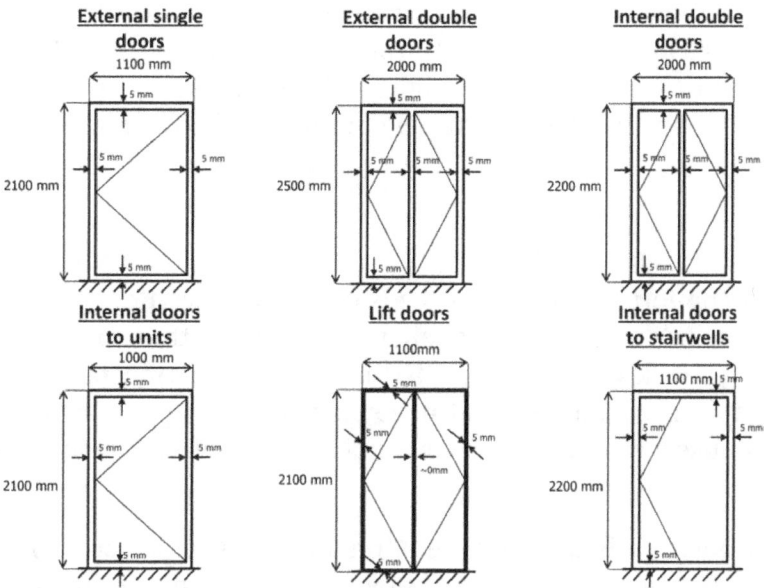

Door specifications

- Modification and control of the design temperature within elevators and stairwell shafts. ssures over the building envelope were calculated following Eurocode.

Working of Stark Effect

In winter, warm air inside a building rises. This pressurizes the top of the building, pushing hot air out and sucking cold air in at the bottom. In summer in an air-conditioned building, stack effect works in reverse because the warmer air is outside the house. Cool inside air tends to fall and get pushed out at the bottom of the building, which draws hot air in at the top.

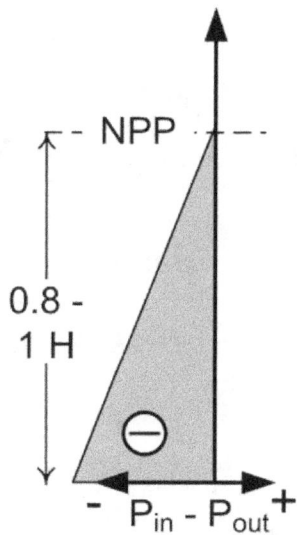

Stack effect is controlled by two things: the height of the building and the difference between inside and outside temperatures. The greater the temperature difference and the taller the building, the greater the pressures created. It's the same principal that creates a strong draft in a chimney.

The pressures, whether positive or negative, are greatest at the bottom and at the top and tend to be neutral somewhere near the middle. In the winter model, the bottom tends to have high negative pressure, the top tends to have high positive pressure, and the middle or "neutral pressure plane," is right in the sweet spot. In summer, negative and positive pressure peaks are reversed.

Importance

Stack effect creates a comfort problem that feeds on itself. In winter, people in the upper floors are overheated, so they open windows. This relieves pressure at the top, which draws cold air in at the bottom, prompting people on lower floors to turn up their thermostats. The problem can really escalate in some multifamily buildings that have poor insulation and air sealing between floors: The overheated penthouse dwellers open their windows, which freezes the feet of the folks at street level.

Stack effect can also cause moisture damage. Moisture rides on air currents, so in any part of a building that experiences a large flow of air between inside and out, moisture will condense on cold surfaces. You can sometimes see the results on brick buildings—as moist air accumulates in the brick, it can cause staining, efflorescence, and spalling from freeze-thaw cycles. But the problems aren't confined to brick. Anytime there is pressure pushing moist inside air—or pulling moist outside air—into the wall cavity, you can definitely get condensation leading to mold and rot.

Energy loss is another effect of the stack. Obviously, when you're heating or cooling inside air, if it escapes, energy is wasted. Again, the problem is cyclical: conditioned air escapes, drawing in unconditioned air that requires more energy to heat or cool it. Rinse and repeat.

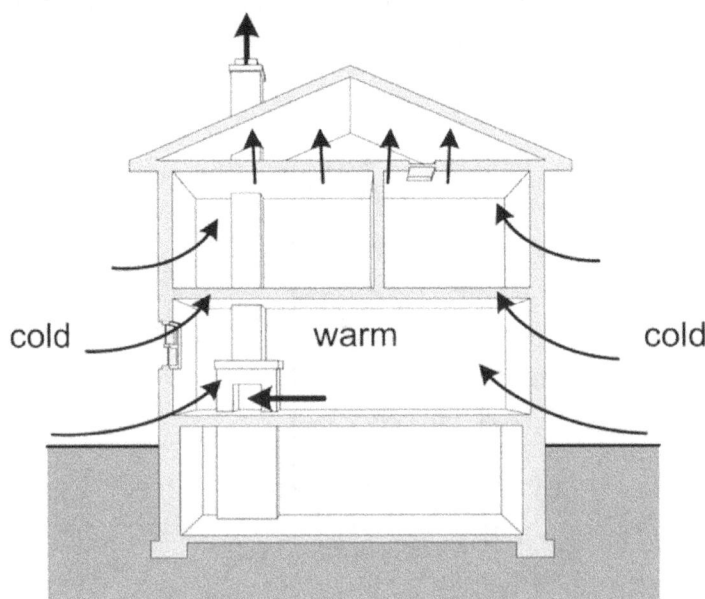

In winter, rising warm air increases pressure at the top of the building, which pushes hot air out at the top and sucks cold air in at the bottom. The reverse occurs in summer, when cool inside air tends to get pushed out at the bottom, which draws hot air in at the top.

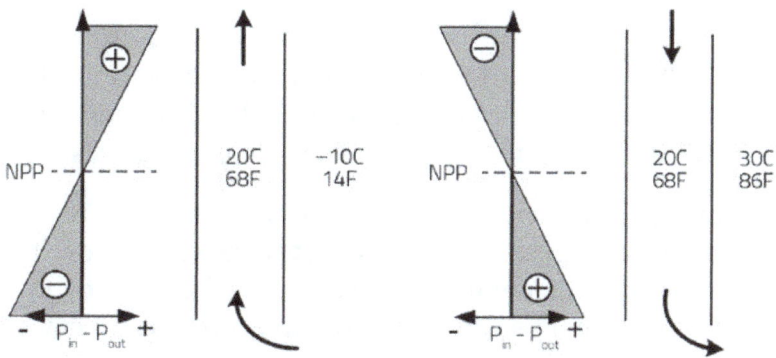

In a heated building (left diagram), warm 68°F air rises, creating positive pressure at the top of a building and negative pressure at the bottom. Warm air escaping to the outside creates an air current that draws cold 14°F air in at the bottom to replace it. The situation is reversed in an air-conditioned building (right diagram), though the temperature differential is typically smaller in summer than in winter. Pressure is equalized at the Neutral Pressure Plane (NPP) in both scenarios, though its location can rise or fall depending on how leaky the building is.

Pros and Cons of Stack Effect Ventilation

PROS

- Does not require the wind to be blowing. Works even when the air is completel.
- Relis on teh natural elements (Temperature and Gravity).
- No Operating Costs.
- No Emissions/No Energy Consumption.

CONS

- Cannot completely control temperatures. The impact depends on exterior temperatures.
- Can bring dust or polluted air into buildings.

Outgassing

Outgassing is the release of a gas that was dissolved, trapped, frozen or absorbed into a material. Outgassing usually happens when components, such as adhesives or other polymers, are exposed to heat.

While some outgassing is normal, it can cause air pollution and even product contamination that can lead to product failures. For example, a computer hard drive is located in a sealed and heated environment. If a polymer were to release plasticizers or other low molecular weight materials in this environment, the outgassed materials could reform in another location of a hard drive and cause shorts, crashes and other malfunctions.

Outgassing Testing

Outgassing testing is also called outgas testing. It uses specialized material testing methods to evaluate products for potential and existing outgassing risks. Common outgassing testing techniques include static headspace analysis and dynamic headspace analysis.

In static headspace analysis, a sample is heated to a specific temperature in a sealed vial. This is done so that volatile compounds can escape into the headspace area above the sample. The gases from the headspace in the vial are then removed and analyzed. This technique is ideal for detecting residual solvents, unreacted monomers and other low molecular weight contaminants.

In dynamic headspace analysis, the sample is sealed in a vessel which is purged with gas while it's being heated. The gas flows through a sorbent material which collects and concentrates the organic constituents being offgassed from the sample. The sorbent material is then desorbed onto the GC/MS for analysis. This method is best used for higher molecular weight contaminants such as siloxanes and plasticizers like phthalates.

Benefits of Outgassing Testing

Outgassing testing ensures products are contaminant free, safe for consumer use and work as they're intended. Outgassing testing can:

- Identify causes or potential causes of product failure or malfunctions.
- Determine if a product is releasing any harmful toxins. (This is often a concern with plastics used for packaging and plastic products such as water bottles or kids' toys).
- Determine if any residual solvents are present.
- Identify strange odors coming from a product.

Outgassing is certainly normal for manufactured products. However, outgassing testing is a must to identify any existing or potential issues, and ensure product quality and safety.

Sensible Heat

Sensible heat is literally the heat that can be felt. It is the energy moving from one system to another that changes the temperaturerather than changing its phase. For example, it warms water rather than melting ice. In other words, it is the heat that can be felt standing near a fire, or standing outside on a sunny day. Sensible heat is used in contrast to latent heat (the heat needed to change from one form of matter to another, which doesn't change temperature), as the two are essentially opposite.

For example, in a cooling system condensation forms due to removal of latent heat, and the refrigerant (cooling liquid) changes temperature due to sensible heat. The sensible heat capacity then describes the capacity required to lower the temperature whereas latent heat capacityis the capacity to remove the moisture from the air.

The amount of sensible heat required is calculated using the following equation:

$$Q = mc\,\Delta T$$

Where :

Q is sensible heat

m is the mass of the object

c is the specific heat of the material

ΔT is the change in temperature

Sensible heat is just the energy associated with temperature change. It will not change the object's phase - whether it is a solid, liquid, or a gas - and will not change the object's pressure or volume. While it is true that increasing the temperature causes the molecules to spread out, the increase caused by sensible heat alone is not significant enough to affect the equation.

Classification and Principles of Storage of Sensible Heat

Thermal energy storage can basically be classified according to the way heat is stored: as sensible heat, in hot liquids and solids, as latent heat in melts and vapour and as chemical heat in chemical compounds. Only the first one is treated here.

Heat - in the physical sense - is a form of energy and can be stored in a variety of ways and for many different applications.

A characteristic property of heat is its temperature: according to this it can be distinguished in low-temperature heat and high-temperature heat. The former is usually applied for domestic hot water supply; it is usually stored in small hot water tanks when used for single family houses, or in large underground containers for large housing projects with hundred and more apartments.

High-temperature heat is applied for all sorts of power processes and in chemical engineering. In these cases, heavily insulated tanks - often pressurised - will be used. There is another criterion for thermal storage, depending on the time involved: short- term storage for a couple of days or long-term storage over a period of months; this is also called seasonal storage. All these terms are found in literature.

Criteria for low- and high-temperature heat are not specified; usually temperatures below 100° C are named low-temperature. A schematic of various concepts of stores for sensible heat is shown in figure below.

The water displacement store has a volume of some hundred litres and is used for hot water supply in houses. The pebble/water reservoir is for the seasonal storage of heat; it is sometimes named "man-made aquifer" and comes in sizes of up to 10.000 m³. The Cowper Regenerator is the device used in steel- and foundry-industry for many years. It consists of a large container filled with ceramic pellets which are alternately heated by flue gases coming from e. g. the blast furnace and cooled by air which is preheated for the furnace.

The solar pond is a special arrangement for an easy store of solar heat. This heat is absorbed at the bottom of the pond and heats the brine there. The salt concentration in the pond increases from top to bottom so that density increases and natural convection is suppressed. Ground Stores,

Multiple Well Stores, Aquifers and Storage Reservoirs are seasonal stores which are supposed to transfer summer heat into the winter demand. The Ruth Accumulator stores hot water under high pressure. When the discharge valve is opened, the pressure is released and the superheated water will evaporate until equilibrium at the corresponding saturation temperature is reached (the water cools down since it provides the evaporation heat to the steam).

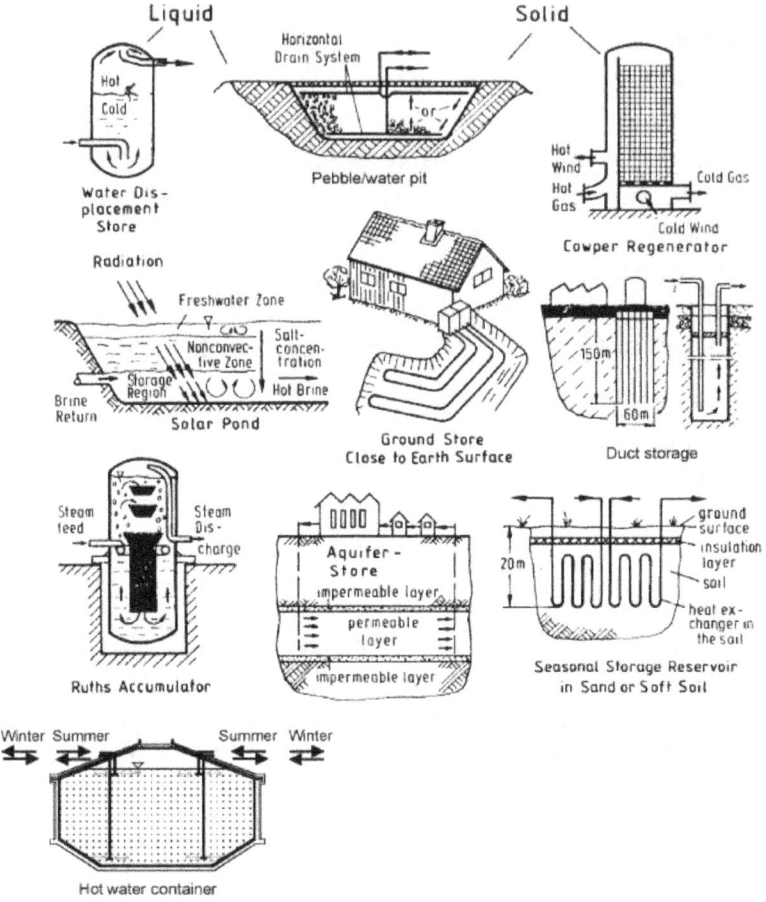

Different Concepts for the Storage of Sensible Heat

An effective store depends very much on the store material. The following requirements should be met:

- Long service life, non-corrosive, non-toxic, non-flammable - large heat storage capacity

- High thermal diffusivity $\alpha = k / \rho c_p$ in e. g. m²/s and heat diffusivity $b = \sqrt{k \rho c_p}$ in $J/m^2 Ks^{1/2}$; with being the thermal conductivity, ρ the density and cp the specific heat capacity.

- Capability to withstand charging/discharging cycles without loss in.performance, store capacity or change in structure.

- Wide availability, simple handling, storage in simple containers.

- Low cost.

The price of the storage medium and the cost of the containment are decisive for the utilisation. Long life and a high cycling stability are pre-requisites for an economic application; i.e., at a price competitive with existing facilities. High temperature diffusivity of the heat storage material provides a quick response to temperature differences; i.e., quick charging and discharging. High heat diffusivity yields a high amount of heat being stored.

Heat transfer processes have to be considered: the heat may be either transferred directly to the storage material as, e. g. in a dry pebble bed with air flows or by way of a heat exchanger as in a solar domestic hot water store where the water- antifreeze mixture flowing through a solar collector has to be separated from the hot water for consumption. Other aspects of selecting a heat store material may be operational advantages in energy supply systems or a larger flexibility in application.

Thermodynamic Considerations

The storage of sensible heat is based thermodynamically speaking on the increase of enthalpy of the material in the store, either a liquid or a solid in most cases. The sensible effect is a change in temperature. The thermal capacity - this is the heat which can be put in the store or withdrawn from it can be obtained by the equation,

$$Q_{12} = m \int_{T_1}^{T_2} c_p(T) dT$$

with the specific heat capacity cp being a function of temperature T, and m mass.

For a temperature independent cp this becomes simply,

$$Q_{12} = mc_p (T_2 - T_1)$$

Where the unit of Q_{12} is, e. g., J. The symbol m stands for the store mass and T_2 denotes the material temperature at the end of the heat absorbing (charging) process and T_1 at the beginning of this process. This heat is released in the respective discharging process.

For pure solids (especially heavy elements), the specific heat per mole of a substance is approximately 3R (Dulong-Petite rule), with R being the molar gas constant ($R=8.31441$ kJ/kmolK). Thus, the molar thermal energy qmol stored in solids can be approximated by,

$$q_{mol} \approx 3R \cdot \Delta T$$

and q_{mol} is measured e. g. in kJ/kmol.

Thus, approximately 25 kJ/kmol can be stored with a temperature difference of $\Delta T = 1K$. With the molar mass M (kg/kmol),

the thermal energy q stored per mass (store capacity in kJ/kg) is obtained,

$$q = c_p \Delta T = q_{mol} / M = 3R \cdot \Delta T / M$$

Heat has a quality, namely its temperature. This determines how much of the heat can be usefully applied according to the second law. This available energy is called Exergy E_x and can be obtained from,

$$E = Q_{12} - \frac{T_{am}}{T} Q_{12}$$

When the Kelvin temperature of the store T = constant during delivery of the heat, with

Tam being the ambient temperature, e.g. 293 K.

For varying temperatures of the store (i.e. stores of a finite size) the exergy change is expressed by,

$$\Delta E_{x12} = (H_2 - H_1) - T_{am}(S_2 - S_1) - (p - p_{am})(V_2 - V_1)$$

with H being enthalpy, S entropy, V volume and p being pressure.

Under atmospheric pressure p = p_{am} and with the 1st and 2nd law this reduces to,

$$\Delta E_{x12} = mc_p \left[(T_2 - T_1) - T_{am} \ln \frac{T_2}{T_1} \right]$$

From above equation it can be observed that exergy changes are not linear with temperature as energy changes.

Exergy is low at low temperatures but increases steeply with an increase in temperature while the energy increase remains the same at low or high temperatures.

A second law evaluation of stores (i.e. by exergy) is more useful than a first law evaluation (by energy). This becomes obvious in *the following example* on the importance of temperature stratification in warm water stores:

Let us assume a container which is half filled with water of 50° C (upper half) and 20° C (lower half). If the two water layers will be fully mixed, the mean temperature is 35° C. The energy in this container remains the same, provided no heat losses occur to the surroundings. Following now equation , assuming an ambient temperature of T = 293° K, a loss in exergy can be calculated. It amounts to x p ΔE 1.23mc . = The energetic consideration does not show any change by the destruction of a thermal layering in the store; the exergetic consideration, however, does indicate the quality of the stored water which, with respect to temperature, has certainly decreased. Half of the container water with 50° C is of more use - e. g. for taking a shower - than all of the water with 35° C.

References

- Swenson, S. Don (1995). HVAC: heating, ventilating, and air conditioning. Homewood, Illinois: American Technical Publishers. ISBN 978-0-8269-0675-5

- How-does-an-hvac-system-work: howardair.com, Retrieved 15 June 2018

- Escombe, A. R.; Oeser, C. C.; Gilman, R. H.; et al. (2007). "Natural ventilation for the prevention of airborne contagion". PLoS Med. 4 (68). doi:10.1371/journal.pmed.0040068

- Use-air-changes-calculation-determine-room-cfm: contractingbusiness.com, Retrieved 31 March 2018

- Dianat, Nazari, I,I. "Characteristic of unintentional carbon monoxide poisoning in Northwest Iran- Tabriz". International Journal of Injury Control and Promotion. Retrieved 2011-11-15

- Thermal-destratification-in-buildings-the-missing-piece-to-the-hvac-puzzle-353736: energy-xprt.com, Retrieved 09 July 2018

- Smolander, J. (2002). "Effect of Cold Exposure on Older Humans". International Journal of Sports Medicine. 23 (2): 86–92. doi:10.1055/s-2002-20137. PMID 11842354

- Understanding-stack-effect: proremodeler.com, Retrieved 29 May 2018

- Al-Kodmany, Kheir (2013). The Future of the City: Tall Buildings and Urban Design. WIT Press. p. 242. ISBN 978-1-84564-410-9

- Sensible-heat-definition-load-equation: study.com, Retrieved 28 April 2018

- Nicol, Fergus; Humphreys, Michael (2002). "Adaptive thermal comfort and sustainable thermal standards for buildings" (PDF). Energy and Buildings. 34 (6): 563–572. doi:10.1016/S0378-7788(02)00006-3

Ventilation and Air Distribution

Ventilation is the introduction of ambient air into a space for the control of indoor air quality by diluting and displacing pollutants from the indoor air setting. This chapter discusses in detail the concepts and principles of airflow, air infiltration and exfiltration, air distribution, displacement ventilation, etc. for an extensive understanding of ventilation and air distribution.

Ventilation

Without proper ventilation, a building can become a gas house of stagnant air, where bacteria and carbon build up making the indoor air hazardous for the inhabitants. In pranayama, the importance of breath and its control is emphasized to imbue health benefits. And, like human bodies, buildings too need to breathe well.

Ventilation is the essential process of replacing stale air with fresh air. Without proper ventilation, buildings become susceptible to stagnant air, where bacteria and carbon make the indoor air more polluted than the air outside.

Sick-building syndrome is common fallout of buildings that do not breathe-that is, the air does not circulate from the inside to outdoors continuously. In poorly ventilated houses, associated with no fresh air supply and its circulation, indoor pollutants and toxins can cause health issues for the inhabitants.

A World Health Organization (WHO) study in 1984 reported that almost 30% of the new and re-modelled construction is plagued by poor indoor air quality.

Given the paucity of land, modern-day construction at times creates stacked building blocks where clusters of flats and offices are built in close proximity, often sacrificing the element of ventilation. Other factors, like extreme temperature led air conditioning, need for insulation, noise levels, pests, and security issues force inhabitants to keep windows and doors tightly closed leading to loss of ventilation opportunities.

Building codes worldwide lay down standards for ventilation. The Indian Building Code (IS: 3362) measures ventilation provided in constructed buildings in the form of the number of 'air changes per hour' or ACH. An air change is the ratio of volume of outside air allowed in the room to the volume of the room in one hour.

This air change is necessary to remove effects of combustion due to cooking, moisture due to water usage, odours in the home, chemicals from printers and computing equipment in offices, and for replacing carbon dioxide with oxygen.

While in living rooms the minimum ventilation standards may require only 3 air changes per hour, this requirement increases to 6 air changes per hour in kitchens due to the heat and fumes produced by cooking.

Good ventilation can be easily incorporated at the building-design stage itself. This helps maintain ambient temperatures indoors without having to resort to energy-guzzling devices like air conditioners. Natural ventilation is the preferred design mode since it conserves valuable energy.

Creation of ventilating windows, keeping these windows on the windward side, door openings, and balconies are meant to aid in natural ventilation. The ideal height of windows is 3.6 feet, as it typically allows wind-based ventilation of homes. Louvres with adjustable angle slats made of wood are a common feature in many buildings, which act as natural ventilation systems; they admit light and fresh air but keep direct sunshine away.

Another natural ventilation system often used in architecture is clerestories (a row of windows in the upper part of the wall). Clerestories are high walls at upper levels of buildings that have many windows above eye level.

This common feature in factory and building complexes helps as warm air rises and escapes through the high windows while cooler, fresher air comes in through the lower windows or vents. From ancient Roman times, atria based design has been widely used as a way to naturally ventilate buildings.

Atriums are large open spaces in the centre of the building. This feature can be seen in large hotels, office complexes, and residential housing and shopping complexes. The central atrium may be enclosed by glass roofing on the top or open to sky. It creates a flow path for air and ventilates it when warm air rises and escape through vents in the upper levels. Atriums also help in optimizing light usage and imparting a light, airy feel to the building.

Air comes into buildings and leaves by three different ways:

- Doors and windows, whenever they are opened.

- Joints, cracks and openings where parts of the building connect, including floors and walls and around windows and pipes.

- Spot ventilation, including fans that pull air from the bathroom.

- Mechanical, whole house systems, to systems in larger buildings that force air into and out of the building.

Challenges Triggered by Poor Ventilation

- When not enough air circulates, pollution builds up indoors. Sometimes efforts to make buildings more energy efficient can backfire by not allowing enough air to move, building up pollution.

- Carbon monoxide can build up to deadly levels indoors without proper ventilation, but it is not the only risk.

- Concentrations of radon, which can cause lung cancer, can increase in homes with low ventilation.

- High humidity outside can make indoor air more humid, increasing the risk of moisture damage indoors, such as mold growth or wood rot.

Possibilities of Ventilation

To guarantee a good indoor air quality it is necessary to remove the "consumed air" (exhaust air) and to supply "fresh air" (supply air).

Pressure differences are the physical reason of each flow and therewith of the air exchange. They can be produced by fans for mechanical ventilation or by natural driving forces for "natural" ventilation.

Natural Ventilation

Natural ventilation has to be distinguished from intended natural ventilation and unintentional natural ventilation. In both cases the driving forces for air exchange are caused by differences in temperature and the wind field. Intended natural ventilation results from:

- *Stack ventilation:* The existing buoyancy inside a gap is used to lead air away *(stack effect)* and to supply the room with fresh air inlets at the same time.

- *Window ventilation:* «shock ventilation" means a short and intensive ventilation through totally open windows; "permanent ventilation" means a long lasting ventilation by tilted windows.

Especially shock ventilation is very effective by opening two opposite windows *(cross ventilation)*. The frequency and duration of ventilation through wind strongly depend on the personal feeling or attitude of the occupants. Therefore, it is very difficult to realise a demand oriented and optimised ventilation through windows because there is either:

- Too much ventilation, which causes high heat losses or

- Too little ventilation, which leads to problems of air quality and damages resulting from humidity.

Unintended natural ventilation (infiltration/exfiltration) results from leakages in the building envelope (joints, leaky windows and doors, installation). In many old buildings the unintended natural ventilation is almost sufficient for the supply with fresh air.

As a consequence, the following disadvantages arise:

- This air exchange is uncontrollable,

- Drafts can arise,

- High ventilation heat losses occur,

- Convective penetration due to humidity damages parts of the building.

Today, the building envelope is implemented as airtight as possible. Using a controlled ventilation will lead to an optimum between good air quality and reduced ventilation heat losses, which are increasingly important as the transmission heat losses are becoming smaller due to the mean while high thermal insulation standard.

Mechanical Ventilation

Fans produce the pressure difference, which is necessary for the air exchange. This makes it possible to control the air exchange and guarantee a hygienically indoor air quality. Mechanical ventilation units are installed in different versions.

Central exhaust ventilation systems are operated with one fan only, which produces a negative pressure in the ventilated section. The supply air flows from outside through leakages or holes into the ventilated section. The exhaust air is blown outside. In order to utilise the heat of the exhaust air, it can be led to heat pumps in bigger systems.

Central ventilation systems have two fans. One fan supplies the building with just the exact quantity of fresh air which the other fan extracts. In buildings, the air inlet and the exhaust openings are not necessarily installed in the same room. Insignificant differences in pressure between supply air (living room, dining room, etc.) and exhaust air section (kitchen, bathroom, etc.) transfer air from the supply zone into the exhaust zone. The warm exhaust air can be used to heat the supply air when the supply air and exhaust air volume rate is directed to one another through a heat exchanger. This recovery is called *recuperative heat recovery*. The different heat exchanger can be distinguished from their air ducts as parallel flow, reverse flow and cross flow heat exchanger. The different types can also be combined, e.g. to a cross reverse flow heat exchanger.

Different characterization of air:

- Outdoor inlet air is called fresh air,
- Air delivered to indoors is called supply air,
- Sucked off indoor air is called extract air,
- Air delivered outdoors is called exhaust air.

Regenerative heat recovery uses the principle of heat storage. The waste heat of the extract air is stored in a buffer and transferred to the supply air later. Central ventilation units of good quality can reach a heat recovery efficiency of 90 %. Today, these central ventilation systems offer the best facilities for supplying a building with fresh air and reducing ventilation heat losses at the same time.

A clever planned duct system, which is required for the already described central ventilation systems, is not necessary for decentralised ventilation units. Therefore, they are suitable for retrofit of buildings or specific ventilation of a single room in which a high pollutant concentration or air humidity can occur. The characterised functions of exhaust air ventilation systems or central ventilation units can also be realised with decentralised ventilation units. Of course decentralised ventilation units do not attain the quite high heat recovery efficiency of central systems. In addition, unsuited exhaust and supply openings can cause short-circuit flow through unfavourable installation.

Normally, the use of mechanical ventilation systems with heat recovery is only reasonable if the building envelope is airtight, so that uncontrolled infiltration, caused by leakage, is very small. If the supply and exhaust airflow rates are badly balanced, the ventilation system puts the building under a pressure difference against outdoor conditions. Then, air may be leaking through the building envelope and increase the ventilation heat losses.

Uses of Ventilation to Protect Health

Use exhaust fans in bathrooms to remove moisture and gases from the house.

Fit your kitchen with an exhaust fan that moves the air to the outside. Use the fan or open a window when cooking to remove fumes and airborne particles.

Make sure gas, propane, wood or other combustion appliances vent completely to the outside. Do not use ventless stoves. Install a carbon monoxide detector in multiple locations in your house.

Vent clothes dryers to the outdoors, too. Clean out the vent regularly to make sure the dust does not block air flow.

If you paint or use hobby supplies or chemicals in your home, add extra ventilation. Open the windows and use a portable window fan to pull the air out of the room.Test your home for radon, and if you have elevated levels, hire a professional to add ventilation to remove it. Radon is the second leading cause of lung cancer.

If the air indoors remains too moist, look for sources of moisture that need to be controlled. If that still does not solve the problem, if a dehumidifier may help. If you use a dehumidifier, make certain you clean it regularly. Check with an air systems specialist to see if your system needs improvements.

Never idle your car in an attached garage. The exhaust can move into your home.

Airflow

Air flow is the rate at which air runs into a purifier in a given period of time. Generally, flow rate is measured in cubic feet per minute (CFM). The primary factor that determines a purifier's air flow rate is the strength of its fans, which are responsible for pulling air into the unit and through the filters. The more "dirty" air a purifier can ingest, the better it will be at sending clean air back into the room.

Air flow can be used to calculate ventilation is the mixing of outside air with inside air. The purpose of the mixing is to keep pollutants and carbon dioxide at the appropriate levels.

Measurement of Air Movement

Smoke Puffer

The Smoke Puffer cannot be used to measure a ventilation rate, but it does give an indication of the direction of air flows, which is another important consideration for proper ventilation. Chemical smoke can be helpful in evaluating HVAC systems, tracking potential contaminant movement, and identifying pressure differentials. Chemical smoke moves from areas of higher pressure to areas of lower pressure if there is an opening between them (e.g., door, utility penetration). Because it is heatless, chemical smoke is extremely sensitive to air currents. Investigators can learn about airflow patterns by observing the direction and speed of smoke movement. Puffs of smoke released at the shell of the building (by doors, windows, or gaps) will indicate whether the HVAC systems are maintaining interior spaces under positive pressure relative to the outdoors.

Safety: The Smoke Puffer uses a reactive chemical to produce smoke. The chemical is an acid and must be used with caution. Avoid inhaling smoke from the Smoke Puffer. Use the Smoke Puffer only under adult supervision.

Wizard Stick

Similar to the Smoke Puffer, except that the Wizard Stick does not use toxic chemicals.

Anemometer or air flow meter (Velocity Measurements)

Airflow in large ductwork can be estimated by measuring air velocity using an anemometer. Measure the air velocity in the ductwork and calculate the outdoor airflow in cubic feet per minute (CFM) at the outdoor air intake of the air handling unit or other convenient location. Additional measurements and calculations are required to get air flow.

Flow Hood

Flow-hoods measure airflow in cubic feet per minute (CFM) at a diffuser or grill. Taking the measurement is simply a matter of holding the hood up to the diffuser and reading the airflow value.

Air Infiltration and Exfiltration

Air infiltration is the movement of air into a building, whereas air exfiltration is the movement of air out of a building. Air leakage into building interiors has a considerable impact on the energy demand of the building. This means that controlling how air moves into and out of buildings is a big part of hose energy efficiency can be improved for buildings. Energy, typically in the form of heat, must be added to or removed from a building in order to maintain a comfortable interior climate for the occupants. Heating, ventilation and air conditioning (HVAC) systems are used to add or remove heat from a building interior and the operation of these systems costs money. Consequently, air infiltration and exfiltration will impact a building's energy costs. Infiltration also affects indoor air quality because it introduces pollutants, allergens, and microbes into the building. It can also result in moisture accumulation in the building envelope because airflow carries with it rainwater and water vapor. Of course all buildings must have sufficient ventilation to be comfortable as well, which is why air infiltration and exfiltration must be handled properly.

Air flows into or out of a building because of pressure differences between the internal and external environments of the building. These pressure differences can be caused by gusting winds or differences in internal and external temperatures. The air will flow through the various openings in the building envelope. These vary from from large openings such as doors and windows, to small cracks and crevices caused by improper installation of envelope components.

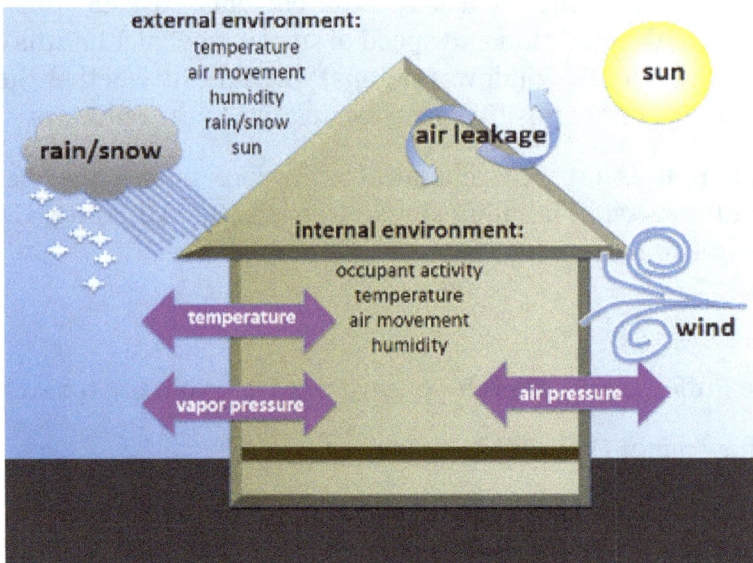

The building envelope helps us control temperature, air pressure and vapour pressure by monitoring the air infiltration and exfiltration

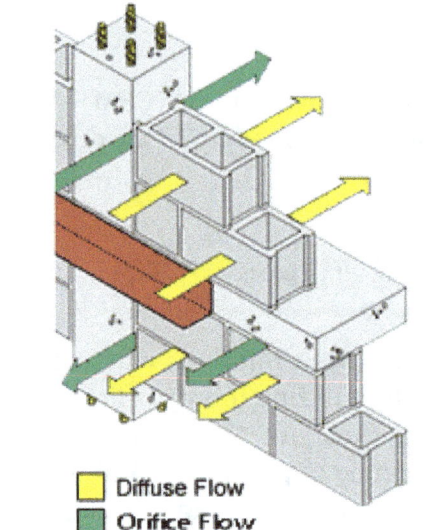

Diffuse and orifice air flow through a building wall

Types and Causes

Air can leak through a building envelope in one of three ways:

- Diffuse flow: occurs when the envelope materials are ineffective in controlling flow. For example the building foundation may be porous. Diffuse flow occurs when the materials has cracks, holes, or they have a high permeability to air. Permeability is the ability of a material to allow liquids and gases to pass through it. Materials with a high permeability to air include fibrous insulation or uncoated concrete blocks.

- Orifice flow: occurs when the path between the air's entry and exits points is linear. For example, air flowing through an open window. This is the green arrow in figure.

- Channel flow: unlike orifice flow, in channel flow the entry and exit points of the air are distant from one another. This allows the air enough time to cool, usually below its dew point, where water vapor will condense into liquid water. This water condensation will damage walls. Channel flow is the most serious type of air leak.

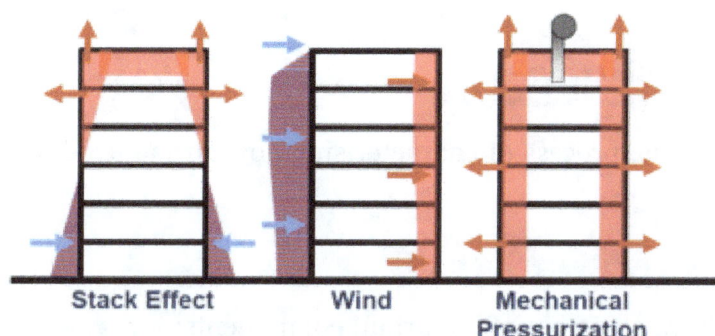

Types of pressure that cause air infiltration/exfiltration (stack pressure, wind pressure, fan pressure)

Differences in air pressure between the interior and exterior of a building will cause infiltration and exfiltration. These pressures include:

- Wind pressure: pressure acting on building walls, roofs, etc. caused by wind gusts.

- Stack pressure: sometimes called chimney effect or buoyancy, stack pressure is caused by a difference in atmospheric pressure at the top and bottom of a building due to differences in height and temperature.

- Fan pressure: pressure caused by mechanical equipment such as an HVAC system. The equipment will typically positively pressurize a building, meaning its interior pressure will be greater than the external pressures acting upon it (wind pressure, atmospheric pressure, etc).

Prevention and Control

Air infiltration and exfiltration in buildings is controlled by an air barrier system. This system includes interconnected materials, flexible sealed joints and building envelope components that create an airtight enclosure. In addition, the air barrier system will also separate conditioned and unconditioned spaces.

The features of an air barrier system that prevent and control infiltration and exfiltration are:

- Continuity: all materials that contribute to controlling infiltration must be properly interconnected to prevent air leakage at any joints. Joints will exist between materials, components, assemblies, and building systems.

- Structural support: each component of the air barrier system must resist any structural loads from wind, stack effect, or fan pressures without coming apart, or being shifted or bent.

- Air impermeability: materials of the air barrier system must not have a high permeability to air. Materials such as fiberboard, perlite board or uncoated concrete blocks should not be used because air passes through them easily.

- Durability: materials chosen for the air barrier system must perform their function for the expected life of the structure or they must be accessible for maintenance.

Room Air Distribution

Room air distribution is the process of characterising how air is introduced to, flows and leaves through various spaces.

Components of Distribution Systems

Distribution Systems have a number of important components:

1. The Air-handling Unit is a cabinet that includes or houses the central furnace, air conditioner, or heat pump and the plenum and blower assembly that forces air through the ductwork.

2. The Supply Ductwork carries air from the air handler to the rooms in a house. Typically each room has at least one supply duct and larger rooms may have several ducts.

3. The Return Ductwork carries air from the conditioned space back to the air handler. Most houses have only one (or two) main return ducts located in a central area.

4. Supply and Return Plenums are boxes made of duct board, metal, drywall or wood that distribute air to individual ducts (or registers).

Supply Duct Systems

Supply ducts deliver air to the spaces that are to be conditioned. The two most common supply duct systems for residences are the trunk and branch system and the radial system because of their versatility, performance, and economy. The spider and perimeter loop systems are other options.

Trunk and Branch System

In the trunk and branch system, a large main supply trunk is connected directly to the air handler or its supply plenum and serves as a supply plenum or an extension to the supply plenum. Smaller branch ducts and runouts are connected to the trunk. The trunk and branch system is adaptable to most houses, but it has more places where leaks can occur. It provides air flows that are easily balanced and can be easily designed to be located inside the conditioned space of the house.

There are several variations of the trunk and branch system. An extended plenum system uses a main supply trunk that is one size and is the simplest and most popular design. The length of the trunk is usually limited to about 24 feet because otherwise the velocity of the air in the trunk gets too low and air flow into branches and runouts close to the air handler becomes poor. Therefore, with a centrally located air handler, this duct system can be installed in homes up to approximately 50 feet long. A reducing plenum system uses a trunk reduction periodically to maintain a more uniform pressure and air velocity in the trunk, which improves air flow in branches and runouts closer to the air handler. Similarly, a reducing trunk system reduces the cross-sectional area of the trunk after every branch duct or runout, but it is the most complex system to design.

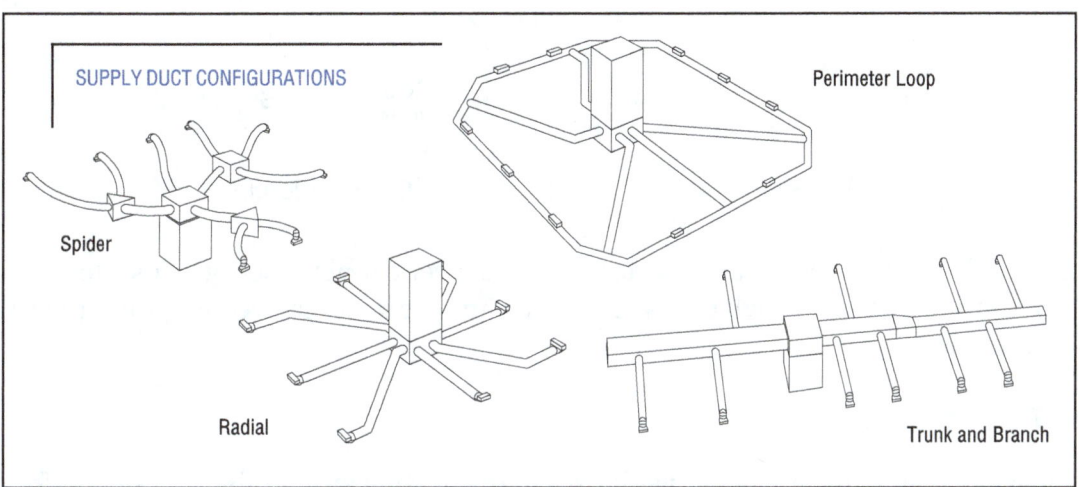

Spider System

A spider system is a more distinct variation of the trunk and branch system. Large supply trunks (usually large-diameter flexible ducts) connect remote mixing boxes to a small, central supply plenum. Smaller branch ducts or runouts take air from the remote mixing boxes to the individual supply outlets. This system is difficult to locate within the conditioned space of the house.

Radial System

In a radial system, there is no main supply trunk; branch ducts or runouts that deliver conditioned air to individual supply outlets are essentially connected directly to the air handler, usually using a small supply plenum. The short, direct duct runs maximize air flow. The radial system is most adaptable to single story homes. Traditionally, this system is associated with an air handler that is centrally located so that ducts are arranged in a radial pattern. However, symmetry is not mandatory, and designs using parallel runouts can be designed so that duct runs remain in the conditioned space (e.g., installed above a dropped ceiling).

Perimeter Loop System

 A perimeter loop system uses a perimeter duct fed from a central supply plenum using several feeder ducts. This system is typically limited to houses built on slab in cold climates and is more difficult to design and install.

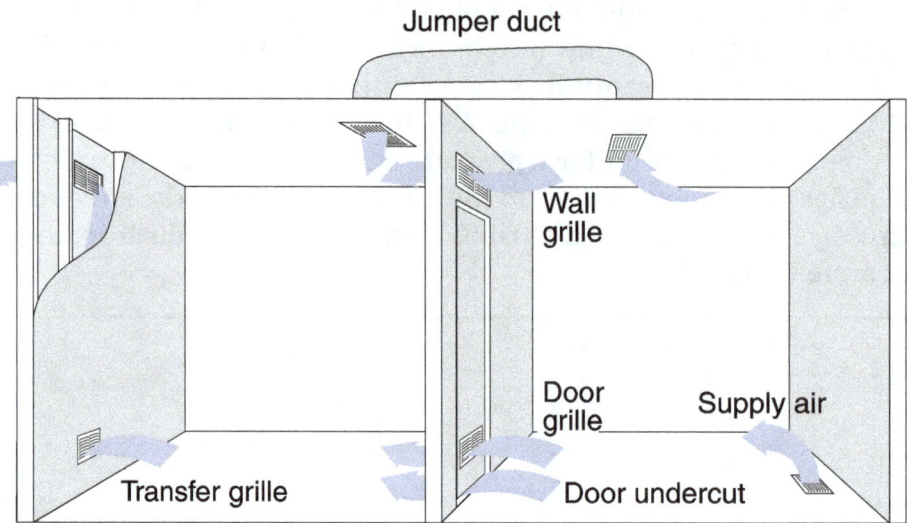

Return Air Techniques: Closed interior doors create a return-air blockage in systems with only one or two returns. Grilles through doors or walls or jumper ducts can reduce house pressures and improve circulation

Return Duct Systems

Return ducts remove room air and deliver it back to the heating and cooling equipment for filtering and reconditioning. Return duct systems are generally classified as either central or multiple-room return.

Multiple -room Return System

A multiple-room return system is designed to return air from each room supplied with conditioned air, especially those that can be isolated from the rest of the house (except bathrooms and perhaps kitchens and mechanical rooms). When properly designed and installed, this is the ultimate return duct system because it ensures that air flow is returned from all rooms (even with doors closed), minimizes pressure imbalances, improves privacy, and is quiet. However, design and installation costs of a multi-room return system are generally higher than costs for a central return system, and higher friction losses can increase blower requirements.

Central Return System

A central return system consists of one or more large grilles located in central areas of the house (e.g., hallway, under stairway) and often close to the air handler. In multi-story houses, a central return is often located on each floor. To ensure proper air flow from all rooms, especially when doors are closed, transfer grilles or jumper ducts must be installed in each room (undercutting interior doors to provide 1 inch of clearance to the floor is usually not sufficient by itself). Transfer grilles are through-the-wall vents that are often located above the interior door frames, although they can be installed in a full wall cavity to reduce noise transmission. The wall cavity must be well sealed to prevent air leakage. Jumper ducts are short ducts routed through the ceiling to minimize noise transfer.

Duct Materials

Air distribution ducts are commonly constructed from sheet metal, rigid fiberglass duct board, or flexible nonmetallic duct. Selection of duct material is based on price, performance, and installation requirements.

Designs that use the house structure or building framing (e.g., building cavities, closets, raised-floor air handler plenums, platform returns, wall stud spaces, panned floor joists) as supply or return ducts can be relatively inexpensive to install. However, they should be avoided because they are difficult to seal and cannot always be insulated. In addition, because such systems tend to be rough and have many twists and turns, it is difficult to design them so as to ensure good air distribution. Even return plenums built under a stairway or in a closet, for example, should be avoided if a completely ducted system is possible.

Sheet Metal

Sheet metal is the most common duct material and can be used on most all supply and return duct applications (for plenums, trunks, branches, and runouts). Sheet metal ducts have a smooth interior surface that offers the least resistance to air flow. When located in an unconditioned space, they must be insulated with either an interior duct liner or exterior insulation. They must also be carefully and completely sealed during construction/installation, using approved tapes or preferably mastic, because each connection, joint, and seam has potential leakage. Screws should be used to mechanically fasten all joints.

Fiberglass Duct Board

Fiberglass duct board is insulated and sealed as part of its construction. It is usually used to form rectangular supply and return trunks, branches, and plenums, although it can be used for runouts as well. Connections should be mechanically fastened using shiplap or V-groove joints and stapling and sealed with pressure-activated tapes and mastic. Fiberglass duct board provides excellent sound attenuation, but its longevity is highly dependent on its closure and fastening systems.

Flexible Nonmetallic Duct

Flexible nonmetallic duct (or flex duct) consists of a duct inner liner supported on the inside by a helix wire coil and covered by blanket insulation with a flexible vapor-barrier jacket on the outside. Flex duct is often used for runouts, with metal collars used to connect the flexible duct to supply plenums, trunks, and branches constructed from sheet metal or duct board. Flex duct is also commonly used as a return duct. Flex duct is factory-insulated and has fewer duct connections and joints. However, these connections and joints must be mechanically fastened using straps and sealed using mastic. Flex duct is easily torn, crushed, pinched, or damaged during installation. It has the highest resistance to air flow. Consequently, if used, it must be properly specified and installed.

Duct and Register Locations

Locating the air handler unit and air distribution system inside the conditioned space of the house is the best way to improve duct system efficiency and is highly recommended. With this design, any duct leakage will be to the inside of the house. It will not significantly affect the energy efficiency of the heating and cooling system because the conditioned air remains inside the house, although air distribution may suffer. Also, ducts located inside the conditioned space need minimal insulation (in hot and humid climates), if any at all. The cost of moving ducts into the conditioned space can be offset by smaller heating and cooling equipment, smaller and less duct work, reduced duct insulation, and lower operating costs.

DUCTS INSIDE CONDITIONED SPACE

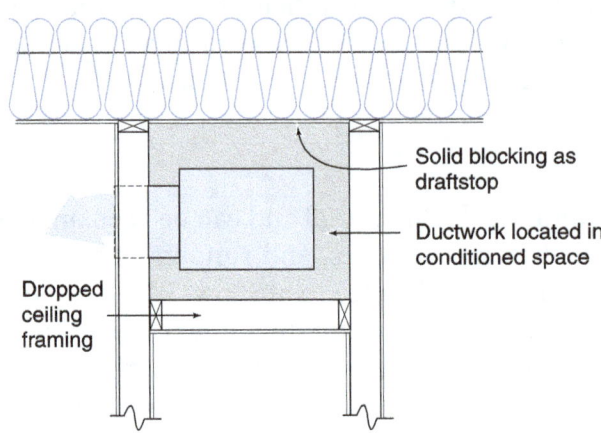

There are several methods for locating ducts inside the conditioned space.

- Place the ducts in a furred-down chase below the ceiling (e.g., dropped ceiling in a hallway), a chase furred-up in the attic, or other such chases. These chases must be specially constructed, air-sealed, and insulated to ensure they are not connected to unconditioned spaces.

- Locate ducts between the floors of a multi-story home (run through the floor trusses or joists). The exterior walls of these floor cavities must be insulated and sealed to ensure they are within the conditioned space. Holes in the cavity for wiring, plumbing, etc., must be sealed to prevent air exchange with unconditioned spaces.

- Locate ducts in a specially-constructed sealed and insulated crawlspace (where the walls of the crawlspace are insulated rather than the ceiling).

Ducts should not be run in exterior walls as a means of moving them into the conditioned space because this reduces the amount of insulation that can be applied to the duct and the wall itself.

A supply outlet is positioned to mix conditioned air with room air and is responsible for most of the air movement within a room. Occupant comfort requires that supply register locations be carefully selected for each room. In cold climates, perimeter floor outlets that blanket portions of the exterior wall (usually windows) with supply air are generally preferred. However, in today's better insulated homes, the need to locate outlets near the perimeter where heat loss occurs is becoming less important. In hot climates, ceiling diffusers or high wall outlets that discharge air parallel to the ceiling are typically installed. In moderate climates, outlet location is less critical. Outlet locations near interior walls can significantly reduce duct lengths (decreasing costs), thermal losses (if ducts are located outside the conditioned space), and blower requirements. To prevent supply air from being swept directly up by kitchen, bathroom, or other exhaust fans, the distance between supply registers and exhaust vents should be kept as large as possible.

The location of the return register has only a secondary effect on room air motion. However, returns can help defeat stratification and improve mixing of room air if they are placed high when cooling is the dominant space-conditioning need and low when heating is dominant. In multi-story homes with both heating and cooling, upper-level returns should be placed high and lower-level returns should be placed low. Otherwise, the location of the return register can be determined by what will minimize duct runs, improve air circulation and mixing of supply air, and impact other considerations such as aesthetics.

Underfloor Air Distribution

Underfloor air distribution (UFAD) is a method of delivering space conditioning in offices and other commercial buildings that is increasingly being considered as a serious alternative to conventional ceiling-based air distribution systems because of the significant benefits that it can provide. This technology uses the open space (underfloor plenum) between the structural concrete slab and the underside of a raised access floor system to deliver conditioned air directly into the occupied zone of the building. Air can be delivered through a variety of supply outlets located at floor level (most common), or as part of the furniture and partitions. UFAD systems have several potential advantages over traditional overhead systems, including improved thermal comfort, improved indoor air quality, and reduced energy use. By combining a building's heating, ventilating, and air-conditioning (HVAC) system with all major power, voice, and data cabling into one easily accessible service plenum under the raised floor, significant improvements can be realized in terms of increased flexibility and reduced costs associated with reconfiguring building services. These raised floor systems are particularly appropriate for office buildings housing today's businesses with their typically extensive use of information technologies and high churn rates.

The approach to HVAC design in commercial buildings has been to supply conditioned air through extensive duct networks to an array of diffusers spaced evenly in the ceiling. conditioned air is both supplied and returned at ceiling level. Ceiling plenums are typically quite large to accommodate the large supply ducts that must fit through them. Return air is most commonly configured as an un-ducted ceiling plenum return. Often referred to as mixing-type air distribution, conventional HVAC systems are designed to promote complete mixing of supply air with room air, thereby maintaining the entire volume of air in the space (floor-to-ceiling) at the desired setpoint temperature and ensuring that an adequate supply of fresh outside air is delivered to the building occupants. This control strategy provides no opportunity to accommodate different thermal preferences among the building occupants or to provide preferential ventilation in the occupied zone.

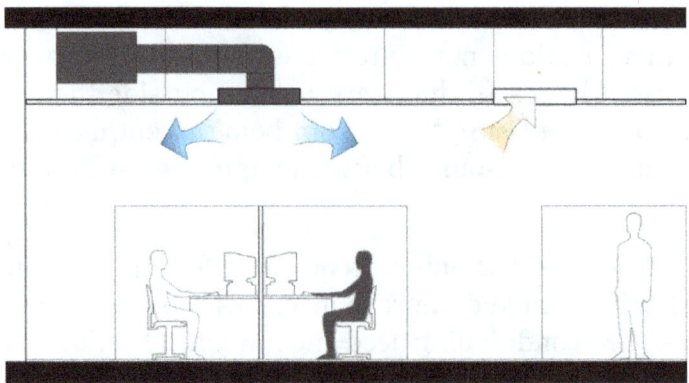

Conventional overhead air distribution system

With UFAD systems, conditioned air from the air handling unit (AHU) is ducted into the under-floor plenum where it typically flows freely to the supply outlets. Underfloor systems are generally configured to have a relatively large number of smaller supply outlets, many in close proximity to the building occupants, as compared to a conventional overhead system. Outlets may be floor diffusers, as shown in figure, or, particularly when part of a task/ambient conditioning (TAC) system, desktop or partition outlets equipped with individual control. If the outlets are adjustable, this arrangement provides an opportunity for nearby occupants to have some amount of control over thermal comfort conditions in their local environment. Air is returned from the room at ceiling level (un-ducted plenum return). This produces an overall floor-to-ceiling air flow pattern that takes advantage of the natural buoyancy produced by heat sources in the office and more efficiently removes heat loads and contaminants from the space, particularly for cooling applications. In contrast to the well-mixed room air conditions of the conventional overhead system, stratification is actually encouraged above head height where increased temperatures and higher levels of pollutants will not affect the occupants.

Underfloor air distribution (UFAD) system

There are three basic approaches to configuring the supply-air side of an UFAD system:

1. Pressurized underfloor plenum with a central air handler delivering air through the plenum and into the space through passive grills/diffusers;

2. Zero-pressure plenum with air delivered to the space through local fan-driven supply outlets in combination with the central air handler;

3. In some arrangements the supply air is ducted through the underfloor plenum to the supply outlets, although in this last configuration certain energy and cost benefits may be reduced compared to the first two approaches.

UFAD Technology Benefits

UFAD systems have several potential advantages over traditional ceiling-based air distribution systems. Well-engineered systems can provide:

1. Improved thermal comfort. By allowing individual workers to have some amount of control over their local thermal environment, individual comfort preferences can be accommodated.

2. Improved ventilation efficiency and indoor air quality. Some improvement in indoor air quality can be achieved by delivering the fresh supply air near the occupant at floor or desktop level, allowing an overall floor-to-ceiling air flow pattern to more efficiently remove contaminants from the occupied zone of the space.

3. Reduced energy use. Energy use can be reduced through a variety of strategies including controlled thermal stratification, higher supply air temperatures, and reduced static pressures in the underfloor plenum.

4. Reduced life cycle building costs. Raised access flooring provides maximum flexibility and significantly lower costs associated with reconfiguring building services.

5. Reduced floor-to-floor height in new construction. UFAD systems can lead to reduced overall service plenum heights compared to conventional overhead systems. A single large overhead plenum to accommodate large supply ducts can be replaced with a smaller ceiling plenum for air return combined with a lower height underfloor plenum for un-ducted air flow and other building services .

6. Improved occupant satisfaction and productivity. Research evidence is mounting that occupant satisfaction and productivity can be increased by giving individuals greater control over their local environment.

These advantages will be realized only if UFAD technology is appropriately designed and applied.

Application Characteristics

UFAD Cooling Load

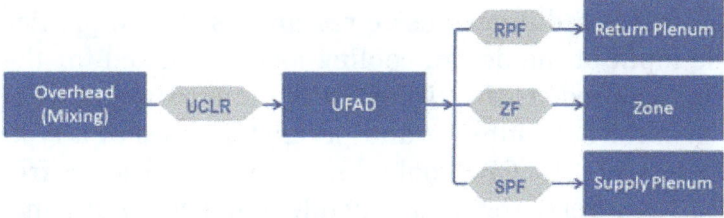

Flow diagram of calculation procedure showing transformation from cooling load calculated for an overhead mixing system into a UFAD cooling load, and then divided between the supply plenum, zone (room), and return plenum.

Cooling load profiles for UFAD systems and overhead systems are different, mainly due to the thermal storage effect of the lighter-weight raised floor panels compared to the heavier mass of a structural floor slab. The mere presence of the raised floor reduces the ability of the slab to store heat, thereby producing for the system with a raised floor higher peak cooling loads compared to the system without a raised floor. In the OH system, particularly in perimeter zones, part of the incoming solar heat gain is stored in the floor slab during the day, thus reducing peak zone cooling loads, and released at night when the system is off. In a UFAD system, the presence of the raised flooring transforms the solar absorbing massive floor slab into a lighter weight material, leading to relatively higher peak zone cooling loads. A modeling study based on EnergyPlus simulations showed that, generally, UFAD has a peak cooling load 19% higher than an overhead cooling load and 22% and 37% of the total zone UFAD cooling load goes to the supply plenum in the perimeter and interior, respectively.

Center for the Built Environment developed a new index UFAD cooling load ratio (UCLR), which is defined by the ratio of the peak cooling load calculated for UFAD to the peak cooling load calculated for a well-mixed system, to calculate the UFAD cooling load for each zone with the traditional peak cooling load of an overhead (well-mixed) system. UCLR is determined by zone type, floor level and the zone orientation. The Supply Plenum Fraction (SPF), Zone Fraction (ZF) and Return Plenum Fraction (RPF) are developed similarly to calculate the supply plenum, zone and return plenum cooling load.

UFAD Design Tools for Zone Airflow Requirements

There are two available design tools for determining zone airflow rate requirements for UFAD system, one is developed at Purdue University as part of the ASHRAE Research Project . The other one is developed at Center for the Built Environment (CBE) at University of California Berkeley.

ASHRAE Research Project developed a simplified tool that predicts the vertical temperature difference between the head and ankle of occupants, the supply air flow rate for one plenum zone, number of diffusers and the air distribution effectiveness. The tool requires users to specify the zone cooling load and the fraction of the cooling load assigned to the underfloor plenum. It also requires users to input the supply air temperature either at the diffuser or at the duct but with the ratio of plenum flowrate to zonal supply flowrate. The tool allows users to select from three type of diffusers and is applicable to seven type of buildings, including office, classroom, workshop, restaurant, retail shop, conference room and auditorium.

The CBE UFAD design tool based on extensive research is able to predict the cooling load for UFAD system with the input of the design cooling load calculated for the same building with an overhead system. It also predicts the airflow rate, room temperature stratification, and the plenum temperature gain for both interior and perimeter zones of a typical multi-story office buildings using UFAD system. The CBE tool allows the user to select from four different plenum configurations (series, reverse series, independent and common) and three floor-diffusers (swirl, square and linear bar grill). An online version of the design tool is publicly available at Center for the Built Environment.

Plenum Air Temperature Rise

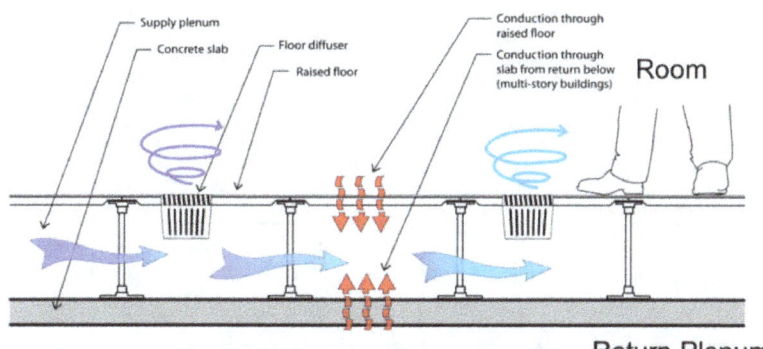

Heat transfer pathways in UFAD system

Plenum supply air temperature rise is the increase of the conditioned air due to convective heat gain as it travels through the underfloor supply plenum from the plenum inlet to the floor diffusers. This phenomenon is also named thermal decay. Plenum air temperature rise is caused by cool supply air coming into contact with warmer than air concrete slab and raised floor. According to a modeling study, air temperature rise can be quite significant (as much as 5° C or 9° F) and subsequently, compared to an idealized simulated UFAD case with no air temperature rise, elevated diffuser air temperatures can lead to higher supply airflow rate and increased fan and chiller energy consumption. The same study found that air temperature rise in summer is higher than in winter and it also depends on the climate. The ground floor with a slab on grade has less temperature rise compared to middle and top floors, and an increase of the supply air temperature causes a decrease in the temperature rise. The temperature rise is not significantly affected by the perimeter zone orientation, the internal heat gain and the window-to-wall ratio. Supply plenum air temperature rise, thus, has implications on the energy saving potential of UFAD systems and their ability to meet cooling requirements with supply temperatures above those of conventional overhead systems. Current research suggests that both energy and thermal performance can be improved in UFAD systems by ducting air to perimeter zones where loads tend to be the greatest. Critics suggest however that such underfloor ducting reduces the benefit of having a low-pressure plenum space, as well as adding design and installation complications when fitting ducts between floor tile pedestals.

Air Leakage in UFAD Plenums

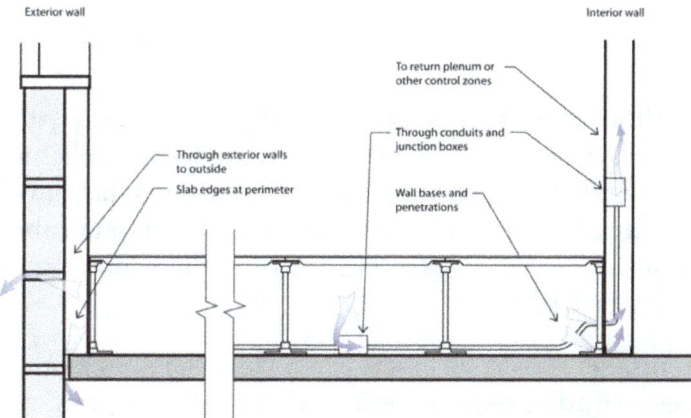

UFAD leakage that does not contribute to cooling, leading to wasted increased fan energy

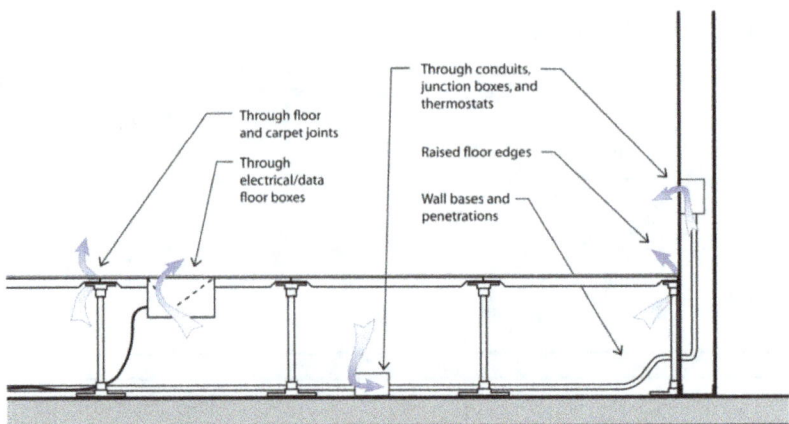

Through conduits, junction boxes, and thermostats

Through floor and carpet joints

Through electrical/data floor boxes

Raised floor edges

Wall bases and penetrations

UFAD leakage into the space, contributing to cooling

Leakage in UFAD supply plenums can be a major cause for inefficiency in a UFAD system. There are two types of leakage—leakage into the space and leakage into pathways that bypass the space.

The first category of leakage does not result in an energy penalty because air is getting to the zone it is intended to cool. The second category of leakage increases fan energy in order to maintain a constant plenum pressure, resulting in increased energy use. Careful consideration needs to be paid in the construction phase of UFAD systems to ensure a well-sealed plenum.

UFAD and Energy

The energy assessment of UFAD systems is a not fully solved issue, which has led to numerous research projects within the building science and mechanical engineering community. Proponents of UFAD point to the lower fan pressures required to deliver air in a building via the plenum as compared to through ducts. Typical plenum pressures are 25 pascals (0.0036 psi) (0.1 inch of water column) or less. The improvements in cooling-system efficiency inherent in operation at higher temperatures save energy, and relatively higher supply air temperatures allow longer periods of economizer operation. However, an economizer strategy is highly climate-dependent and necessitates careful control of humidity to avoid condensation. Critics, on the other hand, cite the shortage of rigorous research and testing to account for variations in climate, system design, thermal comfort and air quality to question whether UFAD is able to deliver improved energy efficiency in practice. Limited simulation tools, the shortage of design standards and relatively scarcity of exemplar projects compound these problems.

Applications

Underfloor air distribution is frequently used in office buildings, particularly highly-reconfigurable and open plan offices where raised floors are desirable for cable management. UFAD is also common in command centers, IT data centers and Server rooms that have large cooling loads from electronic equipment and requirements for routing power and data cables. The ASHRAE Underfloor Air Distribution Design Guide suggests that any building considering a raised floor for cable distribution should consider UFAD.

Specific space considerations should be taken when using UFAD systems in laboratories because of its critical room pressurization requirements and potential migration of chemicals into the access floor plenum due to spillage. UFAD systems are not recommended in some specific facilities or spaces, such as small non-residential buildings, wet spaces like restrooms and pool areas, kitchens and dining areas and gymnasiums, because UFAD may result in especially difficult or costly in design. UFAD systems may also be used with other HVAC systems, like displacement ventilation, overhead air distribution systems, radiant ceiling or chilled beam systems to get better performance.

UFAD Compared to other Distribution Systems

Overhead

Conventional *overhead mixing systems* usually locate both the supply and return air ducts at the ceiling level. Supply air is supplied at velocities higher than typically acceptable for human comfort and the air temperature may be lower, higher, or the same as desired room temperature depending on the cooling/heating load. High-speed turbulent air jets incoming supply air mix with the room air.

A well-engineered UFAD systems have several potential advantages over traditional overhead systems, such as layout flexibility, improved thermal comfort, improved ventilation efficiency and indoor air quality, improved energy efficiency in suitable climates and reduced life cycle costs.

Displacement Ventilation systems (DV) work on similar principals as UFAD systems. DV systems deliver cool air into the conditioned space at or near the floor level and return air at the ceiling level. This works by utilizing the natural buoyancy of warm air and the thermal plumes generated by heat sources as cooler air is delivered from lower elevations. While similar, UFAD tends to encourage more mixing within the occupied zone. The major practical differences are that in UFAD, air is supplied at a higher velocity through smaller-size supply outlets than in DV, and the supply outlets are usually controlled by the occupants.

Displacement Ventilation

In systems using the concept of displacement ventilation (DV) the air is supplied at a lower temperature in relation to the room air. The air is usually supplied at floor level, directly in the occupied zone.

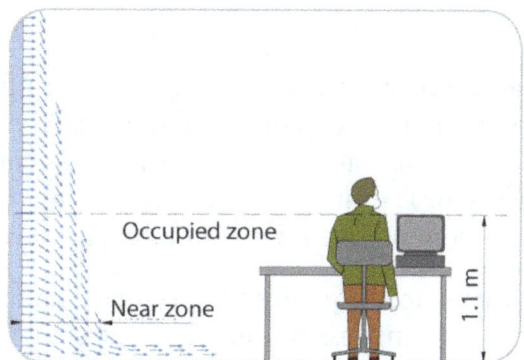

This creates a stratified flow where the cooled supply air flows out into the room under the warmer room air. The flow in the room is based on natural air movements where the air is driven by a difference in density and by convection flows from heat-releasing activities and processes.

Convection flows at heat sources generate a vertical air flow in the room, thereby creating a clean zone on the bottom and a polluted zone on top.

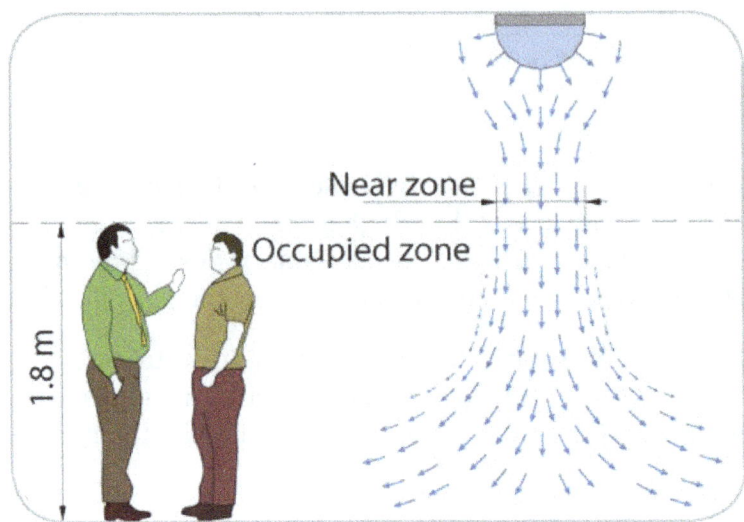

A high level of heat activity from heat sources generates bigger convection flows, resulting in the air rising more strongly and greater entrainment of the air around the source. Textile ducting can create displacement ventilation even when fitted at high level because there is no mixing of the flows.

In general, displacement ventilation systems operate at both low velocity and very low pressure. This delivers an immediate additional benefit of low operational noise combined with a minimal draught within the occupied area.

In the temperature gradient associated with displacement ventilation, the ceiling-level air temperature is higher than that of the occupied zone. The first effect of this is that the supply temperature can be higher than that of traditional mixing systems. This higher air temperature means longer spells of free-cooling, reducing energy use through the lifetime of the system. A higher return air temperature also renders the system ideal for use with an air handling unit (AHU). There is a consequent dramatic reduction in the heating energy needed to meet room conditions after the energy recovery device.

Supplying air near the floor level rather than at the ceiling has significant design implications. To avoid creating uncomfortable drafts for occupants situated near the diffusers, DV supplies air at higher temperatures—above 63° F—and at a lower velocity. To achieve adequate cooling under these conditions, the air diffusers for DV must be much larger than those of conventional systems. These larger diffusers must be accommodated in the building layout.

The lower velocity of air delivery reduces pressure requirements and allows fans to run more slowly, consuming less energy and producing less noise. This makes DV an excellent choice when lower noise levels are desired. Also, because the supply air temperature is higher than a traditional mixed-flow air system, the economizer cycle is used well into the summer season.

Designing for Variations

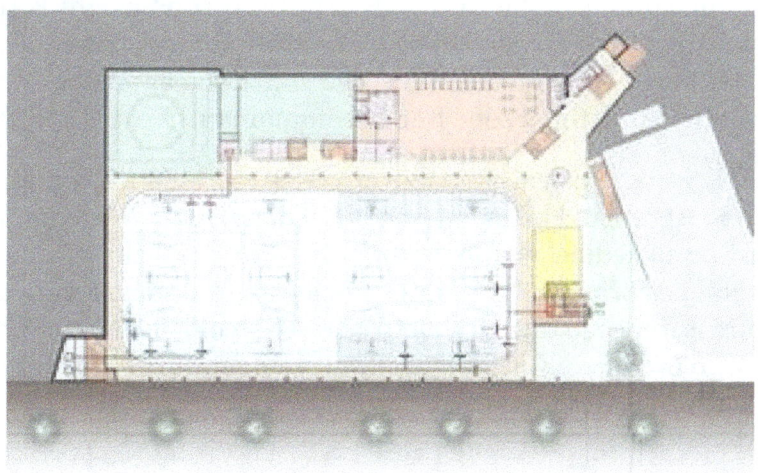

When applying DV, consider the ventilation rate and cooling load density to ensure that the cooling requirements align with the supply air system's ability to offset the heat gain. In spaces where the internal heat gain density exceeds 15 Btuh/sq ft, take extra care to ensure comfort. DV can be supplemented with localized cooling terminals such as chilled cooling panels to address higher heat gain densities. Emissions from equipment or food preparation should be captured at their source and removed, and care must be taken to limit downdrafts driven by cold walls or cooled ceiling panels, which may lead to higher contaminant concentrations.

A noticeable vertical temperature gradient sets up between the floor and the ceiling. This vertically stratified temperature field needs to be stable for DV to function properly. The temperature in the space is expected to vary linearly with space height in the stratified zone and is nearly constant in horizontal directions, except for regions near diffusers. However, it is difficult to predict the temperature gradient because some of the influential parameters are difficult to account for. These parameters include the radiation between the ceiling and the floor, the variable heat output of equipment and people, and the location of people within the space. A small temperature difference between wall and room air can result in noticeable downdrafts with cold walls and updrafts with warm walls. As the load in the space varies, the buoyancy forces that drive airflow will also vary. Computational fluid dynamics analysis can be used to model the complex air and temperature patterns and predict temperature gradients in the occupied zone.

Because drafts and wide temperature stratification may reduce occupant comfort, the air temperature near the floor and the vertical temperature gradient in occupied zones are the most important parameters in evaluating DV. To reduce the temperature gradient, the supply flow rate must be increased, but this could lead to high air velocity at floor level and high draft risk, as well as consume more energy. To offset heating loads in the winter and augment vertical airflow, perimeter heating terminals such as radiant panels or fin tube can be used. Convective heating terminals can boost the buoyancy-driven airflow within the space.

When heat removal is the main objective, a temperature-based design can be used to match the supply airflow rate and temperature to the current load in the space. When contaminants are a major concern, a shift zone method is used. The shift zone is the boundary between the lower, non-recirculating zone and the upper zone, which has recirculation flow. The concentration of contaminants is at its maximum at the shift zone. The shift zone height is the height above the floor at which the total amount of air carried in convective plumes above a heat source is equal to the supply airflow distributed through diffusers. The shift zone approach is popularly applied in industry to keep the human breathing zone free of contaminants.

Relative humidity (RH) is also critical. RH above 70% is too high for comfort and may result in fungal growth. At 74° F, RH should be maintained at 50% to 60%. To accomplish this, some or all of the supply air must be cooled below room dew point. In humid climates, return air bypass improves dehumidification performance by routing outside air and some return air through the cooling coil where moisture is removed from the airstream. This cold and dry air is then mixed with the conditioned return air to produce a warmer supply air temperature with lower moisture content.

Equipment Needs

Air handling unit discharge temperatures are higher for DV than for mixed flow. DV works best with a centralized plant using air handling units and chilled water designed to provide 62 to 67° F supply air to the space. Using this temperature range, free cooling may be available most of the year, and the overcooling and reheat associated with mixed-flow systems is eliminated. The actual energy consumption will depend on the control strategies and air handling systems used. Enthalpy recovery devices can be applied to reduce the dehumidification load of the cooling process.

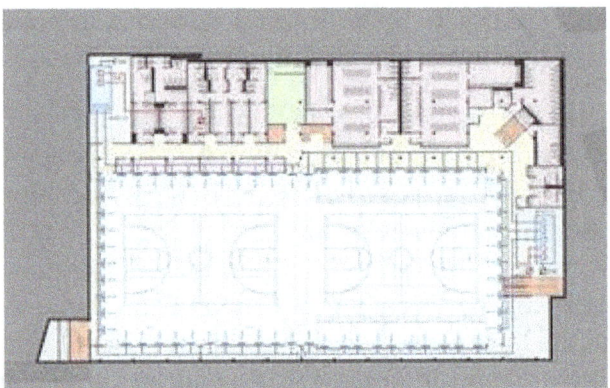

DV adds complexity to supply air ducting because of the integration of low supply air devices within the programmed space. The architectural design needs to incorporate these large air devices. Generally, supply air can be released from wall ducts that run under windows. As an alternative, air can be supplied through floor plenums, or ceiling jets can be used to send a vertical column of air to the floor and provide personal temperature control. Offices wider than 20 ft may need more than one supply air device. Low sidewall floor diffusers integrated into column enclosures can be effective for large open-plan spaces. Diffuser location and coverage needs to be researched and carefully planned.

Fewer diffusers and less ductwork can be expected, with the understanding that DV diffusers are more expensive than conventional mixed-flow diffusers. Select supply air diffusers with a discharge velocity lower than 50 fpm, as lower supply velocities reduce drafts and noise. DV's limited cooling capacity restricts usage to applications with relatively low heat gain. It may be found that DV is ideal for core zones but not appropriate for perimeter zones because the cooling load is too high.

If DV is used in perimeter spaces, a separate heating system may be needed to maintain airflow patterns. Convectors, baseboard, and radiant panels are effective but come with additional cost.

Diffuser type is key to providing the higher air volume rates needed for higher heat loads. Displacement outlets may be located almost anywhere in a room, but the conventional approach has been to locate them at or near floor level. They should be located to take advantage of naturally occurring thermal stratification within the room. Volume control dampers can be located in branch ducts supplying diffusers.

To avoid drafts near diffusers, diffuser performance is critical, and the velocity in the occupied zone, especially near diffusers, must be well controlled, with a velocity no higher than 50 fpm. The entrainment of room air will decrease the temperature gradient in the occupied zone. Blending supply air quickly will reduce drafts. Information on diffuser aspiration and modulation can be found in product catalogs. For diffusers, the distance from a wall-mounted diffuser to a 40 fpm velocity contour along the center line is an important parameter for comfort.

Flow from several diffusers placed close together on the wall will merge to a two-dimensional flow, resulting in velocity lower than a radial-flow single diffuser. If diffusers have slanting discharges and are too close together, the supply airflows can meet and project straight into the room for several feet.

Applications

Typically, ceilings should be higher than 9 ft for the displacement effect to take place. Higher ceiling heights enable DV to remove larger heat loads. DV should not be used if high temperatures above 7 ft are unacceptable. It blends well with vaulted ceilings, daylighting, and radiant floor heating. DV may also be useful in core spaces where cooling is always needed. However, perimeter zones with high cooling loads need a close look. The supply airflow rate should be the higher of the required ventilation airflow rate and the supply airflow rate based on cooling load.

Compliance with ASHRAE 55 ensures that the supply temperature, velocity, and vertical temperature gradient in the occupied zone are acceptable. It limits the magnitude of supply room temperature difference and space cooling loads for a given supply airflow rate.

A number of design parameters must be considered, including supply airflow rate, supply air temperature, air temperature at floor level, vertical temperature gradient, maximum air velocity at floor level, stratification height, contaminant concentration gradient, energy consumption, first cost, and maintenance. It is prudent to be cautious and clarify assumptions when applying recommendations from design guides.

DV Pros and Cons

The unique characteristics of large underground institutional facilities can be addressed by applying technologies that are merging into common practice. Displacement ventilation (DV) offers many advantages, including improved indoor air quality, low acoustics, and energy efficiency. It is now being applied in large open spaces as well as offices and classrooms. If a building's design accommodates the special requirements of DV, it can be a healthy option and energy-efficient opportunity.

Advantages

1. DV provides better acoustics and better air quality than mixed-flow systems. Mixed-flow systems tend to be louder because of the higher velocity required from diffusers. Diffuser noise can be difficult to attenuate. Applying DV diffusers rather than mixed-flow diffusers can reduce sound levels by an NC factor of 5.

2. Lower supply velocity at diffusers means lower pressure drop, smaller fans, and less energy consumption. Fan horsepower reductions can be attributed to less air movement.

3. DV can use fewer diffusers and less ductwork.

4. DV introduces supply air at the playing/teaching floor, improving indoor air quality by reducing accumulation of CO_2, odor, and indoor contaminants.

5. DV has a higher ventilation effectiveness than mixed-flow systems.

6. Free cooling may be available most of the year.

7. If 100% outdoor air and exhaust is used, the heat gain due to the lights and roof can be eliminated from building cooling loads.

Disadvantages

1. DV cannot be applied as widely as mixed-air systems.

2. DV can add complexity to supply air ducting.

3. DV diffusers are more expensive than mixed-flow diffusers.

4. The room neutral temperature for a DV system is higher than that of a conventional mixing system.

Natural Ventilation

Natural ventilation is the process of pulling fresh air into a building from the outside. In turn, this fresh air helps force the warm, dirty air inside of the building out through the opening in the roof. This becomes done, without mechanical assistance. While that is the definition, in a nutshell, it's actual application and use is much more complex.

There are essentially two types of ventilation that occur naturally in buildings. There is either wind-based ventilation (also known as cross ventilation) or Buoyancy-driven ventilation (also known as stack-effect ventilation.

Wind-based Ventilation (Cross-ventilation)

Wind moving past a building will create different areas of high and low pressure. The windward side of the building is an area of high pressure while the leeward side & roof are low pressure. Strategically placing different sizes and types of openings in the high and low-pressure areas cause air to move through the building at an increased rate and in the desired direction.

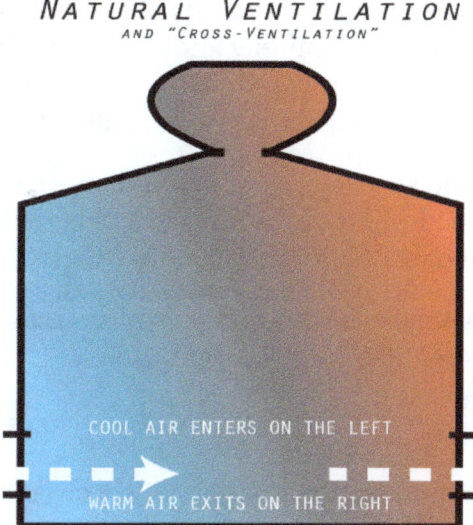

The easiest way to understand wind-based or wind-driven ventilation is to look at an open window. Fresh air comes in through one side, moves throughout the building, and then goes out through the opposite side, pushing the stagnant warm air that was in the building along with it. Additionally this is a great, low-cost ventilation solution in certain facilities.

However, it should be noted that wind-based ventilation isn't always ideal in areas where there's a lot of pollution or dust. It also doesn't work in areas where the wind is not favorable, i.e. wind blows in the opposite direction and has no effect. Wind-based ventilation can also cause a building to particularly draughty, which can be problematic in the winter months as well.

Buoyancy-driven Ventilation

It seems complicated, but it works basically like a fireplace. Because warm air rises and cool air stays low, this process forces warm air up.

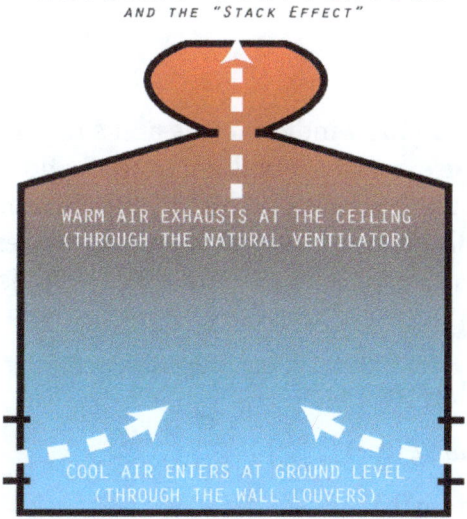

Heat generated within buildings rises towards the ceiling. In large buildings, where the structure is of significant height, the natural tendency of warm air to rise generates air movement throughout the entire building. This is what is commonly known as thermal buoyancy or the stack effect. Warm air rises through the structure and escapes via the natural ventilator. The wall louvers at the floor level allow cool air to enter, replacing the warm air that escapes.

The tall roofs of industrial buildings create substantial differences in temperatures and pressures, which is why industrial sites are ideal locations for gravity ventilation systems. Exploit these differences to direct air up and out of the building naturally.

Unlike cross-ventilation, this method can still work on cool and calm days. For natural ventilation, all year round buoyancy-driven ventilation is the right choice.

Combined Ventilation

Natural Ventilators can also be used in combination with powered ventilation to keep the entire building comfortable throughout the day.

Powered ventilation is the only option in some spaces. Facilities that use chemicals or have lower ceilings height or just don't produce quite enough heat require powered ventilation. Ultimately, there are a lot of reasons why powered ventilation may be necessary. More and more plant managers are going with systems that use a natural solution for larger areas, and powered ventilation for work rooms and smaller areas throughout their facilities.

Types of Natural Ventilation Effects

Wind can blow air through openings in the wall on the windward side of the building, and suck air out of openings on the leeward side and the roof. Temperature differences between warm air inside and cool air outside can cause the air in the room to rise and exit at the ceiling or ridge, and enter via lower openings in the wall. Similarly, buoyancy caused by differences in humidity can allow a pressurized column of dense, evaporatively cooled air to supply a space, and lighter, warmer, humid air to exhaust near the top. These three types of natural ventilation effects are further described below.

Wind

Wind causes a positive pressure on the windward side and a negative pressure on the leeward side of buildings. To equalize pressure, fresh air will enter any windward opening and be exhausted from any leeward opening. In summer, wind is used to supply as much fresh air as possible while in winter, ventilation is normally reduced to levels sufficient to remove excess moisture and pollutants. An expression for the volume of airflow induced by wind is:

Qwind = K x A x V, where

Qwind = volume of airflow (m^3/h)

A = area of smaller opening (m^2)

V = outdoor wind speed (m/h)

K = coefficient of effectiveness

The coefficient of effectiveness depends on the angle of the wind and the relative size of entry and exit openings. It ranges from about 0.4 for wind hitting an opening at a 45° angle of incidence to 0.8 for wind hitting directly at a 90° angle.

Sometimes wind flow prevails parallel to a building wall rather than perpendicular to it. In this case it is still possible to induce wind ventilation by architectural features or by the way a casement window opens. For example, if the wind blows from east to west along a north-facing wall, the first window (which opens out) would have hinges on the left-hand side to act as a scoop and direct wind into the room. The second window would hinge on the right-hand side so the opening is down-wind from the open glass pane and the negative pressure draws air out of the room.

It is important to avoid obstructions between the windward inlets and leeward exhaust openings. Avoid partitions in a room oriented perpendicular to the airflow. On the other hand, accepted design avoids inlet and outlet windows directly across from each other (you shouldn't be able to see through the building, in one window and out the other), in order to promote more mixing and improve the effectiveness of the ventilation.

Buoyancy

Buoyancy ventilation may be temperature-induced (stack ventilation) or humidity induced (cool tower). The two can be combined by having a cool tower deliver evaporatively cooled air low in a space, and then rely on the increased buoyancy of the humid air as it warms to exhaust air from the space through a stack. The cool air supply to the space is pressurized by weight of the column of cool air above it. Although both cool towers and stacks have been used separately, the author feels that cool towers should only be used in conjunction with stack ventilation of the space in order to ensure stability of the flow. Buoyancy results from the difference in air density. The density of air depends on temperature and humidity (cool air is heavier than warm air at the same humidity and dry air is heavier than humid air at the same temperature). Within the cool tower itself the effect of temperature and humidity are pulling in opposite directions (temperature down, humidity up). Within the room, heat and humidity given off by occupants and other internal sources both tend to make air rise. The stale, heated air escapes from openings in the ceiling or roof and permits fresh air to enter lower openings to replace it. Stack effect ventilation is an especially effective strategy in winter, when indoor/outdoor temperature difference is at a maximum. Stack effect ventilation will not work in summer (wind or humidity drivers would be preferred) because it requires that the indoors be warmer than outdoors, an undesirable situation in summer. A chimney heated by solar energy can be used to drive the stack effect without increasing room temperature, and solar chimneys are very widely used to ventilate composting toilets in parks.

An expression for the airflow induced by the stack effect is:

Qstack = Cd*A*$[2gh(T_i - T_o)/T_i]^{1/2}$, where

Qstack = volume of ventilation rate (m³/s)

Cd = 0.65, a discharge coefficient.

A = free area of inlet opening (m²), which equals area of outlet opening.

g = 9.8 (m/s²). the acceleration due to gravity

h = vertical distance between inlet and outlet midpoints (m)

Ti = average temperature of indoor air (K), note that 27° C = 300 K.

To = average temperature of outdoor air (K).

Cool tower ventilation is only effective where outdoor humidity is very low.

The total airflow due to natural ventilation results from the combined pressure effects of wind, buoyancy caused by temperature and humidity, plus any other effects from sources such as fans.

Design Recommendations

The specific approach and design of natural ventilation systems will vary based on building type and local climate. However, the amount of ventilation depends critically on the careful design of internal spaces, and the size and placement of openings in the building.

- Maximize wind-induced ventilation by siting the ridge of a building perpendicular to the summer winds.

  ◦ Approximate wind directions are summarized in seasonal "wind rose" diagrams available from the National Oceanographic and Atmospheric Administration (NOAA). However, these roses are usually based on data taken at airports; actual values at a remote building site can differ dramatically.

  ◦ Buildings should be sited where summer wind obstructions are minimal. A windbreak of evergreen trees may also be useful to mitigate cold winter winds that tend to come predominantly from the north.

- Naturally ventilated buildings should be narrow.

  ◦ It is difficult to distribute fresh air to all portions of a very wide building using natural ventilation. The maximum width that one could expect to ventilate naturally is estimated at 45 ft. Consequently, buildings that rely on natural ventilation often have an articulated floor plan.

- Each room should have two separate supply and exhaust openings. Locate exhaust high above inlet to maximize stack effect. Orient windows across the room and offset from each other to maximize mixing within the room while minimizing the obstructions to airflow within the room.

- Window openings should be operable by the occupants.

- Provide ridge vents.

  ◦ A ridge vent is an opening at the highest point in the roof that offers a good outlet for both buoyancy and wind-induced ventilation. The ridge opening should be free of obstructions to allow air to freely flow out of the building.

- Allow for adequate internal airflow.

  ◦ In addition to the primary consideration of airflow in and out of the building, airflow between the rooms of the building is important. When possible, interior doors should be designed to be open to encourage whole-building ventilation. If privacy is required, ventilation can be provided through high louvers or transoms.

- Consider the use of clerestories or vented skylights.

    ◦ A clerestory or a vented skylight will provide an opening for stale air to escape in a buoyancy ventilation strategy. The light well of the skylight could also act as a solar chimney to augment the flow. Openings lower in the structure, such as basement windows, must be provided to complete the ventilation system.

- Provide attic ventilation.

    ◦ In buildings with attics, ventilating the attic space greatly reduces heat transfer to conditioned rooms below. Ventilated attics are about 30°F cooler than unventilated attics.

- Consider the use of fan-assisted cooling strategies.

    ◦ Ceiling and whole-building fans can provide up to 9°F effective temperature drop at one tenth the electrical energy consumption of mechanical air-conditioning systems.

- Determine if the building will benefit from an open- or closed-building ventilation approach.

    ◦ A closed-building approach works well in hot, dry climates where there is a large variation in temperature from day to night. A massive building is ventilated at night, then, closed in the morning to keep out the hot daytime air. Occupants are then cooled by radiant exchange with the massive walls and floor.

    ◦ An open-building approach works well in warm and humid areas, where the temperature does not change much from day to night. In this case, daytime cross-ventilation is encouraged to maintain indoor temperatures close to outdoor temperatures.

- Use mechanical cooling in hot, humid climates.

- Try to allow natural ventilation to cool the mass of the building at night in hot climates.

- Open staircases provide stack effect ventilation, but observe all fire and smoke precautions for enclosed stairways.

visitor center at Zion National Park showing downdraft cooling tower with evaporative media at the top, and exhaust through high clerestory windows

Natural ventilation in most climates will not move interior conditions into the comfort zone 100% of the time. Make sure the building occupants understand that 3% to 5% of the time thermal comfort may not be achieved. This makes natural ventilation most appropriate for buildings where space conditioning is not expected. As a designer it is important to understand the challenge of simultaneously designing for natural ventilation and mechanical cooling—it can be difficult to design structures that are intended to rely on both natural ventilation and artificial cooling. A naturally ventilated structure often includes an articulated plan and large window and door openings, while an artificially conditioned building is sometimes best served by a compact plan with sealed windows. Moreover, interpret wind data carefully. Local topography, vegetation, and surrounding buildings have an effect on the speed of wind hitting a building. Wind data collected at airports may not tell you very much about local microclimate conditions that can be heavily influenced by natural and man-made obstructions. Hints about what type of natural ventilation strategies might be most effective can often be found in a region's historic and vernacular construction practices.

Materials and Methods of Construction

Some of the materials and methods used to design proper natural ventilation systems in buildings are solar chimneys, wind towers, and summer ventilation control methods. A solar chimney may be an effective solution where prevailing breezes are not dependable enough to rely on wind-induced ventilation and where keeping indoor temperature sufficiently above outdoor temperature to drive buoyant flow would be unacceptably warm. The chimney is isolated from the occupied space and can be heated as much as possible by the sun or other means. Air is simply exhausted out the top of the chimney creating suction at the bottom which is used to extract stale air.

Wind towers, often topped with fabric sails that direct wind into the building, are a common feature in historic Arabic architecture, and are known as "malqafs." The incoming air is often routed past a fountain to achieve evaporative cooling as well as ventilation. At night, the process is reversed and the wind tower acts as a chimney to vent room air. A modern variation called a "Cool Tower" puts evaporative cooling elements at the top of the tower to pressurize the supply air with cool, dense air.

In the summer, when the outside temperature is below the desired inside temperature, windows should be opened to maximize fresh air intake. Lots of airflow is needed to maintain the inside temperature at no more than 3-5 °F above the outside temperature. During hot, calm days, air exchange rates will be very low and the tendency will be for inside temperatures to rise above the outside temperature. The use of fan-forced ventilation or thermal mass for radiant cooling may be important in controlling these maximum temperatures.

Natural Ventination and Chimney

Natural ventilation relies on the wind and the "chimney effect" to keep a home cool. Natural ventilation works best in climates with cool nights and regular breezes.

The wind will naturally ventilate your home by entering or leaving windows, depending on their orientation to the wind. When wind blows against your home, air is forced into your windows on the side facing into the wind, while a natural vacuum effect tends to draw air out of windows on the leeward (downwind) side. In coastal climates, many seaside buildings are designed with large ocean-facing windows to take advantage of cooling sea breezes. For drier climates, natural ventilation involves avoiding heat buildup during the day and ventilating at night.

The chimney effect relies on convection and occurs when cool air enters a home on the first floor or basement, absorbs heat in the room, rises, and exits through upstairs windows. This creates a partial vacuum, which pulls more air in through lower-level windows. The effect works best in open-air designs with cathedral ceilings and windows located near the top of the house, in clerestories, or in operable skylights.

Mixed-mode Ventilation

Mixed mode ventilation (hybrid ventilation) is a combination of natural and mechanical ventilation and exploits both systems as required. This entails either interchangeable periods of natural ventilation and mechanical ventilation, or a balancing of two principles such that mechanical ventilation takes over when the external conditions so require. The result is an effective and energy-efficient ventilation solution that maintains healthy indoor climate and comfort. In order to use mixed mode ventilation solution conditions for the use of natural ventilation must be in place.

Mixed mode ventilation solutions, like all other solutions that utilise natural ventilation, are based on the principle of provinding healthy indoor climate and comfort, delivered with minimal energy consumption and at minimal cost.

Mixed mode ventilation can either be manually operated, or automatically controlled. A manually operated mixed ventilation scheme might be as simple in its operation as turning on an exhaust fan at the same time as you open windows for cross ventilation.

Automated systems are usually a fair bit more complex. These types of systems use sensors that recognise when natural ventilation is not doing a good enough job, and accordingly switch over to powered ventilation. If the outdoor wind speed and temperature were to change again, the powered ventilation would switch off and the system would revert to natural ventilation.

Types of Mixed Mode System

1. Concurrent *(Same space, same time)*

   Concurrent mixed-mode operation is the most prevalent design strategy in practice today, in which the air-conditioning system and operable windows operate in the same space and at the same time. The HVAC system may serve as supplemental or "background" ventilation and cooling while occupants are free to open windows based on individual preference. Typical examples include open-plan office space with standard VAV air-conditioning systems and operable windows, where perhaps perimeter VAV zones may go to minimum air when sensor indicates that a window has been opened.

2. Change-over *(Same space, different times)*

    Change-over designs are becoming increasingly common, where the building "changes-over" between natural ventilation and air-conditioning on a seasonal or even daily basis. The building automation system may determine the mode of operating based on outdoor temperature, an occupancy sensor, a window (open or closed) sensor, or based on operator commands. Typical examples include individual offices with operable windows and personal air conditioning units that shut down for a given office anytime a sensor indicates that a window has been opened; or a building envelope where automatic louvers open to provide natural ventilation when the HVAC system is in economizer mode, and then close when the system is in cooling or heating mode.

3. Zoned *(Different spaces, same time)*

    Zoned systems are also common, where different zones within the building have different conditioning strategies. Typical examples include naturally ventilated office buildings with operable windows and a ducted heating/ventilation system, or supplemental mechanical cooling provided only to conference rooms. For many mixed-mode buildings, operating conditions sometimes deviate somewhat from their original design intent (e.g., a building originally designed for seasonal changeover between air-conditioning and natural ventilation may, in practice, operate both systems concurrently).

Advantages/Disadvantages

-   No energy is required to provide ventilation or cooling where the building operates a natural ventilation strategy only.

-   For the Sydney climate, mixed mode is more practical than natural ventilation except in spaces that have limited comfort requirements.

-   Opening windows have higher infiltration than sealed windows, which can increase heating and cooling loads in extreme conditions.

-   Management of opening windows in open plan areas is difficult due to wind effects and different perceptions of comfort for different occupants.

-   Open windows and louvres can permit external noise and pollutants into the space, especially in an urban setting.

-   Natural ventilation is difficult to provide to deep plan buildings

Energy Efficiency

A very low energy solution when no active cooling is provided.

A mixed mode strategy will also be a low energy solution compared to conventional systems, although actual savings will depend on how frequently the mechanical cooling system is used instead of natural ventilation.

Running Costs

The cooling effect from opening a window is 'free'. If windows are open and the mechanical cooling or heating system is also running then this will waste energy. Effective building management will help reduce running costs in this situation e.g. if the system is automatically controlled by a building management system (BMS).

Retrofit/Improvement Opportunities

The major areas for improvement for natural ventilation and mixed mode systems are:

- Measures to eliminate windows being open when the building is actively heating or cooling.
- Optimisation of the control changeover between natural ventilation and active conditioning.
- Review and improvement of airflows during natural ventilation mode.
- Upgrade of controls for boilers, chillers and associated pumps.

Control improvements can be implemented with the tenants in-situ. Replacing non-openable windows with openable windows to provide natural ventilation is not a cost effective solution to saving energy. If, however, a building has been provided with openable windows in addition to an air conditioning system, that system may be turned off at some points during the year to save energy/cost and become a mixed mode building.

Applicable Buildings

Natural ventilation is possible for light industrial spaces, but most other spaces will use mixed mode as active cooling is required on hot summer days.

Natural ventilation and mixed mode operation is typically not suited to buildings with surrounding pollution issues e.g. city centres.

Floor Plate Implications

Opened windows on a single elevation will only ventilate to a depth of circa 7m, which increases to 15m (including allowance for 1m wide central walkway across floor plate) if windows are on opposing elevations. Spaces deeper than 15m need a more complex ventilation strategy which typically requires an open central atrium to achieve the required stack effect.

Temperature Control/Occupant Comfort

Natural ventilation is generally unsuitable for buildings with a high cooling demand or in city centres where noise and pollution make opening windows undesirable. Windows provide limited control of how much air enters a space, which will differ depending on external conditions. For example, air flow will fluctuate depending on whether it is a still or windy day. Natural ventilation is difficult to implement in open plan floor plates owing to the conflicting perceptions and preferences of occupants, and may cause occupant dissatisfaction as a result.

Maintenance Implications

Manual windows will require occasional adjustment to ensure they are in a fully operable condition. Automatic window/louvre opening devices will require more maintenance.

Identification

There will be no or very few air grilles and no mechanical plant where the building operates a natural ventilation strategy. The majority of windows will be openable. There may be a horizontal transfer grille for the air to leave a room or floor area and enter an atrium. A building operating a mixed mode strategy will feature an air conditioning system as well as openings such as windows and louvres.

Demand Controlled Ventilation

In a Demand Controlled Ventilation (DCV) system the ventilation airflow rate is continuously matched with the actual demand. By this, the DCV system offers an obvious advantage compared to conventional Constant Air Volume flow (CAV) systems. Due to decreased average airflow rates, less energy is needed for fan operation and for heating and cooling of the supply air. Moreover, a DCV system based on room temperature control also eliminates the need of additional heating in rooms when the cooling capacity of the supply air exceeds the cooling capacity needed, e.g. when the room is unoccupied or when the solar heat load is low. This advantage in terms of energy savings is often overlooked.

A DCV system based on air-quality control adapts the airflow rate to the actual pollution load, which often is proportional to the occupancy. For example, all the rooms in an office building or in a school are almost never occupied at the same time. Furthermore, it is most unlikely that the peak level of occupancy occurs simultaneously in all occupied rooms. For instance, quite a few studies show that in cellular offices usually less than about 50% of the office rooms are occupied at the same time. Studies in a number of school buildings reveal the average occupancy to be up to 30%. The bigger the variation between the minimum and peak loads, the more energy savings can be expected with a DCV system based on air-quality control.

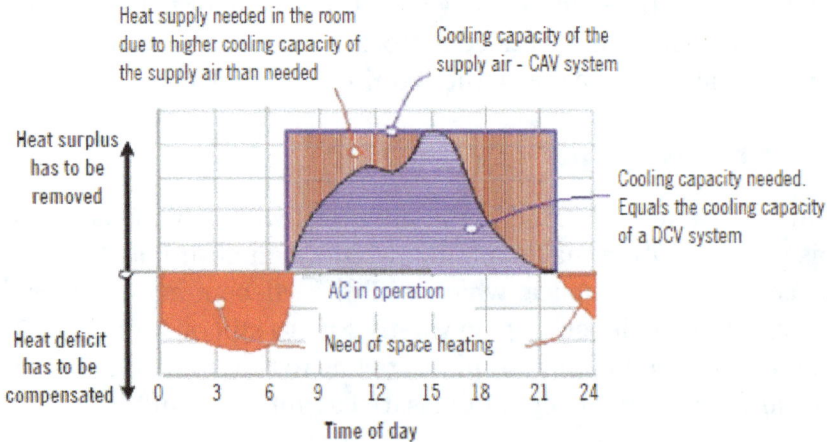

Comparison of DCV system with CAV system in the application of thermal comfort control (air-based cooling). Typical variations of heat surplus and heat deficit in an office room. In a CAV system additional heat supply is needed in the room when internal heat generation falls below the cooling capacity of the supply air.

Even if there are obvious functional advantages with a DCV system, the system itself is somewhat more complex than a CAV system. A competent design and a careful installation, commissioning and maintenance are required to ensure the expected performance. DCV systems have been in use for 30 years or more, but in quite limited numbers due to poor experiences. Most common problems have been poor indoor climate and operational problems. The main reason has been lack of proper components and limited understanding of DCV system design. Improper selection and design of airflow control and supply air devices has been a common cause of excessive noise and draught in occupied spaces, but has also led to under- or over-cooling of the premises. Furthermore, quite a few problems with performance of the sensors controlling the airflow rates have been reported.

The main indicator for thermal comfort is room temperature or sometimes a combination of temperature and humidity (specific enthalpy).

The main indicator for air quality is the composition of air in terms of gases, particles, etc. Carbon dioxide is the most common indicator for air quality related to human occupancy. The ventilation flow rate, as determined by requirements on air composition or air quality, is known as the hygienic ventilation flow rate. Control may rely on the measured state of air (feed-back control), the measured load (feedforward or predictive control) or a combination of these two.

According to this concept, a DCV system requires a VAV system, and most VAV systems, but not all, operate as DCV systems. Only the VAV systems where the airflow rate varies continuously according to the actual demand are considered as DCV systems. Thus, VAV systems where the airflow rate varies according to a predetermined pattern or manual control are not considered as DCV systems.

Technical Challenges of Building up a DCV System

The fundamental requirement on a DCV system is to assure good indoor climate with reference to indoor air quality, thermal comfort and acoustic environment. In addition, this goal should be achieved cost-effectively and with a minimum of purchased energy.

The complexity of the technical solution of a DCV system and the system layout is to a great extent dependent on how the pressure unbalance in the system is absorbed. The airflow rate in individual rooms is adapted to the demand either by variable supply air diffusers in the room or by airflow control dampers in the duct in connection with the supply air devices. As the variation of airflow rates leads to variations of the static pressure in the system, pressure control methods should be applied to avoid an excessive increase in pressure at the airflow control devices when the average airflow rate is low. A common practice for static pressure control is to adjust the fan speed, while the static pressure in the system is kept on a level that can assure a proper operation of the flow control devices.

Depending on the location of the static pressure sensor and on the variation range of the airflow rates, the pressure rise that must be throttled off somewhere in the air distribution system can still become relatively high. It is essential that this pressure variation is managed in a way that good functioning properties of the airflow control devices are ensured. This means that the airflow rates supplied to the individual rooms must be insensitive to the pressure variations in the duct. In practice it is often necessary for the chosen airflow control components to manage a pressure drop of at least 100 Pa without a disturbing generation of noise. It is also vital that the supply air diffusers provide a stable air movement pattern in the room, which must be independent of the supply airflow rate. Neither at high air flow rates, nor low airflow rates, should there be any risk of draught in the occupied zone. The risk of draught may occur when traditional CAV supply air devices are used in a DCV system.

For energy efficient performance of a DCV system, the airflow control devices and supply air diffusers should be able to control the airflow rate within a wide range. In order to obtain an efficient airflow rate control in a temperature controlled DCV system, it is important that the supplied air has a low temperature; about +15° C to +16° C can be recommended. The supply air devices must be able to supply air with such comparably low air temperature without any risk of draught. This is one reason why supply air devices for displacement ventilation are unsuitable in temperature controlled DCV systems. A displacement type of supply air device usually needs about at least +190C supply air temperature and that defines the supply air temperature for the whole system. Therewith the supply air will have a very limited cooling capacity and the DCV system will in practice operate as a CAV system. One displacement supply air device might jeopardize the DCV function of the whole system.

With low supply air temperatures it is also important that the duct system manages the varying airflow rates without considerable heat gains. When the airflow rates are decreased the heat gains in the duct system can have significant effect on the supply air temperature. Insulating all of the ducts in the system is a basic requirement to maintain the required cooling capacity of the supply air. Bigger effect for decreasing the heat gains can be achieved by increasing the insulation thickness on the main ducts instead of increasing the insulation thickness on the connection ducts.

It should also be noted that a DCV system puts a higher demand on the fan design and its performance. The fan is expected to operate in a stable manner over a wider airflow range compared with a CAV system. If this is not addressed, problems with noise may occur in addition to problems with controlling the airflow rates accurately, leading to poor energy performance of the fans in general.

Control of Room Air Quality – a Challenge for the Sensing Technology

An air quality controlled DCV system adapts the airflow rate continuously to the actual pollutant emissions from activities and processes in the room. However, control of indoor air quality with sensor methods can be somewhat more complicated than control of thermal comfort.

Firstly, it can be difficult to define the reference parameters influencing indoor air quality that the sensors must measure. There are no sensors that measure the "quality" of air. Instead, quantitative parameters, as the composition of air in terms of gases and particles, can be measured and linked to the air quality. However, in many cases the link between the perception of air quality, the concentration levels of various substances and their influence on comfort and health is still somewhat unclear.

Secondly, the available sensing technologies set the limits. The choice of an indicator or pollutant for control of air quality is to a great extent dependent of the possibilities to measure this parameter. The sensors applicable for indoor air quality control are based on measurement of gaseous compounds. It is relatively simple to measure just one substance, as carbon dioxide (CO_2), which is considered as a fairly good indicator of pollutants due to human occupancy. Greater challenges are faced when a combination of substances, e.g. volatile organic compounds are to be considered. There is a large amount of different organic compounds at different concentrations in indoor air. The real health and comfort impact of many of these individual components and their combinations at the usually low concentrations is still relatively unclear. The mixed-gas sensors available on the market measure non-selectively a wide range of gases and they do not indicate which gases are detected or what their concentration might be.

Thirdly, even if there are available technologies for measuring the required parameter, the sensor must fulfil certain requirements in order to be applicable for ventilation control. Sensors for indoor air quality control with a DCV system should give fast, stable and reliable output signals corresponding to the value of the specified quantity measured. Incorrect measurement of specified reference quantities can lead to under- or over-ventilated rooms, resulting in uncomfortable indoor climate or excessive use of energy. In addition, the correct location of sensors is vital to achieve the required performance from any control system.

Quantitative requirements for indoor air quality sensors have been developed based on ventilation guidelines and standards. A detailed sensor study with a number of CO_2-sensors and mixedgas sensors showed that there are several CO_2- sensors that fulfil the established requirements set on sensors. However, the application of the tested mixed-gas sensors for ventilation control is not decided upon. It is not clear how the output of mixed-gas sensors should be interpreted. Another limitation for the application of mixed-gas sensors is related to lack of available standards describing acceptable concentrations for many common air contaminants for non-industrial buildings.

Examples of Case Studies on DCV System Performance

Demand controlled ventilation systems have been installed in several different types of commercial buildings in Sweden. However, only few studies on energy and indoor climate performance have been reported recently.

Two case studies were carried out about 5 years ago in two office buildings where one type of a modern DCV system solution was installed. The first case study was a retrofit project, where the existing CAV ventilation system was replaced with a DCV system. The second case study building consists of two parts: one part was newly built and the other part fully renovated. Both of the case study buildings have similar DCV system layout. The DCV supply air devices in the rooms are controlling the airflow rates after temperature and occupancy sensors. The supply air temperature is about +15° C.

The case studies showed that it is possible to build-up a well-functioning DCV system that functions as expected from thermal comfort and energy points of view. Due to the low supply air temperature, the airflow rate control versus the heat load in the rooms is effective. The air-toair heat recovery system accounts for all almost all the air heating needed and there is almost no need for additional heating with the heating coil in any of the case study buildings.

The data were measured hourly during one year in one of the building parts of the second case study building. This building part consists of 58 cell office rooms and a number of meeting rooms. The total designed airflow rate for all rooms is 3.0 m3/s. As shown in the figure, the DCV system never reached this airflow rate during the measurement period. The maximum measured supply airflow rate was approximately 76% of the maximum airflow rate of all rooms. Moreover, it operated with less than 45% of the design airflow rate during 80% of the operating hours, i.e. during about 3900 hours/year. These low values compared to the design values can be explained by a low use of the rooms. In figure comparison is made with a CAV system, with the values estimated as if the system would have operated with constant air volume flow rate.

Taking into account occupancy profiles and load variations in the building, there is a possibility to optimise the size of the central components and the duct system of a DCV system. Additionally, dimensioning based on the airflow rates needed in reality is important for achieving good control properties of the whole DCV system. Overdimensioning of fans, when the main operating range will remain at a very low capacity, can lead to control problems and considerable decrease in total efficiency of the fan system. Improved knowledge of the occupancy patterns in different types of buildings would be of great help for a correct DCV system design.

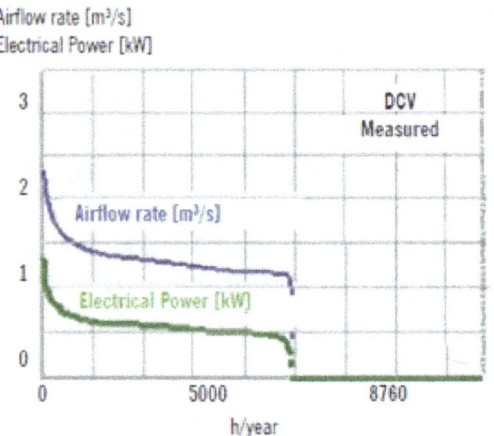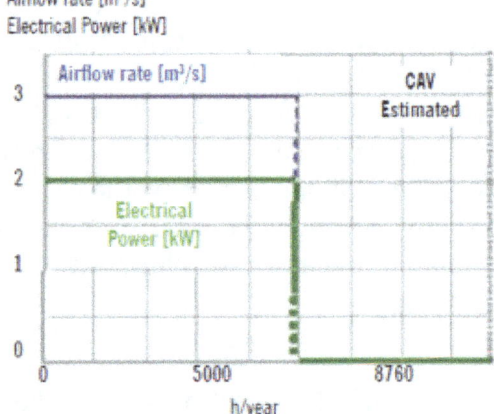

Measured supply airflow rate and supply air fan electrical power in a 2500 m² office building with a DCV system. One year hourly measurements (left graph) compared with a corresponding theoretical (reference) CAV-system (right graph). The total designed airflow rate for the building is 3.0 m³/s

How to assure that the expected performance of a DCV system is achieved?

In order to guarantee adequate performance of a DCV system and to achieve the expected energy savings, it must be designed, installed, commissioned and operated under a constant and complete commissioning process. The following key-points should be kept in mind in order to build up a well-functioning DCV system:

- Full understanding of the system principles and accurate system design. It is essential to correctly understand the concept and function of the whole system, from individual rooms to central air handling unit.

- Right choice of components. It is essential to clarify which requirements are to be set on the system and its components. A detailed analysis should be carried out about which demand decides the DCV function, which the proper indicators of this demand are, which sensing technologies are applicable, which suitable components for airflow rate control are to be chosen, how the pressure control is to be obtained, etc.

- Full understanding of the control system for DCV systems, right choice of control strategies and components. It is essential to evaluate which requirements to set on the control system and which input parameters to choose for the whole system. The locations of the sensing points are equally important.

- Correct installation and commissioning. It is important that the system is installed according to its design and that it is commissioned in a way that ensures that right operation of the system is achieved.

- Proper maintenance of the system. The personnel taking care of the system must be fully qualified to manage the complex control and fault detection.

Getting the best out of a DCV system implies that all of the above mentioned factors are taken into account. More detailed practical guidelines in these areas should be developed in order to assure the best performance of demand controlled ventilation.

Energy Recovery Ventilation

An Energy Recovery Ventilation system uses a heat exchanger to not only transfer sensible heat (basic temperature difference) but also latent heat (the heat required to covert a solid state into a liquid or vapour). Because both temperature and moisture are transferred, ERVs are considered total enthalpic devices. The core material for ERVs is permeable to enable the transfer of moisture.

One of the benefits of using an ERV is that it transfers heat and moisture between incoming and outgoing airstreams. This means it takes much less energy to heat/cool the air you bring into your home, ultimately saving you money on heating and cooling costs.

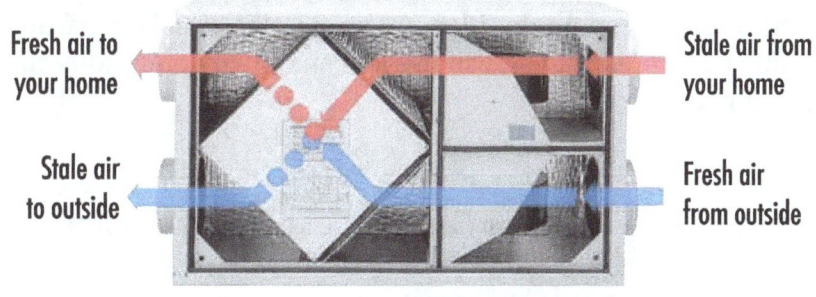

Fresh air to your home

Stale air to outside

Stale air from your home

Fresh air from outside

Wintertime example

- During the winter: As your stale, hot indoor air is expelled, it preheats the fresh, cold outdoor air that's coming in. You will use much less energy to heat the air once it gets inside.

- During the summer: As your stale, conditioned indoor air is expelled, it removes heat from the fresh, hot air that's coming in. This cools the incoming air and reduces the energy needed to cool it to a comfortable level inside.

Science behind Energy Recovery Ventilation

Heat Transfer

So how exactly does this heat transfer take place? Basically, a simple transfer of energy occurs in the Energy Max Transfer Core, using enthalpic technology based on the second law of thermodynamics. This law essentially states that energy will spontaneously move from areas of high energy to areas of lower energy.

What we know as "temperature" is actually a measure of the average kinetic energy of molecules in a substance. When two substances with different temperatures meet, energy (i.e. heat) is transferred from the high temperature substance to the low temperature substance.

The transfer of energy is never complete, but if given enough time, heat will be transferred to the lower temperature substance until the two have equal temperatures. It's similar to mixing one cup of hot water and one cup of cold water to make two cups of warm water.

Moisture Transfer

Energy recovery ventilators transfer moisture similarly to how they transfer heat. An ERV heat exchanger is usually made of a different material, often paper, so that moisture can transfer between incoming and outgoing air. Like temperature, moisture in the air will move from an area of high concentration to an area of low concentration. This is how the Energy Max Transfer Core works in the Aprilaire Model 8100 Energy Recovery Ventilation System (ERV).

- During the winter:

  As it's expelled, your warm, humidified indoor air transfers moisture to the cold, and typically dry, outdoor air that's coming in. Therefore, you will use less energy to humidify the air once it gets inside.

- During the summer:

  As it's expelled, your stale, dehumidified indoor air steals moisture from the fresh, humid air that's coming in. Therefore, you will use less energy to dehumidify the air to a comfortable level once it gets inside.

References

- ASHRAE Technical Resource Group On Underfloor Air Design (2013). UFAD GUIDE Design, Construction and Operation of Underfloor Air Distribution Systems. W. Stephen Comstock. ISBN 978-1-936504-49-7.

- Ventilation-buildings-breathe, our-initiatives, healthy-air: lung.org, Retrieved 19 June 2018

- Kwang Ho, Lee; Stefano Schiavon; Fred Bauman; Tom Webster (2012). "Thermal decay in underfloor air distribution (UFAD) systems: Fundamentals and influence on system performance". Applied Energy. 91 (1): 197–207. doi:10.1016/j.apenergy.2011.09.011.

- What-is-displacement-ventilation: ke-fibertec.com, Retrieved 15 July 2018

- Lehrer, David et. al (2003), Reality new research findings on underfloor air distribution systems., Center for the Built Environment, UC Berkeley, p. 6, retrieved 2011-11-29

- When-where-to-use-displacement-ventilation: csemag.com, Retrieved 25 May 2018

- Woods, James (2004), "What real world experience tells us about the UFAD alternative", ASHRAE Journal, retrieved 2011-11-29

- Natural-ventilation-mixed-mode-systems: betterbuildingspartnership.com.au, Retrieved 11 May 2018

3

Central Heating Systems

A central heating system is used to provide heat and warmth to the interiors of a building from one point to multiple rooms. All the diverse aspects of central heating systems such as geothermal heat pump, air source heat pump, ground-coupled heat exchanger, space heater, underfloor heating, etc. have been carefully analyzed in this chapter.

Central Heating

Central Heating is the even distribution of heat to every room in your home from a central point or heat source.

The warm water distributes heat evenly and thoroughly to every area of your home via a network of pipes. The pipe network can be connected to radiators in each room or can be looped under the floor for underfloor heating, both systems releasing a gentle heat into your home.

- Radiators: Usually require high temperatures that can only currently be achieved with combusting boilers.

- Under Floor Heating: Uses lower temperature water and can be heated by any heat source.

The term central heating covers hydronic heating systems with a central boiler or furnace either inside the building being heated or in the immediate vicinity.

Heat is generated in the boiler. Pipes carry the heated water to the building's heat sources (radiators) and return the cooled water to the boiler again.

Originally, many central heating systems were designed to be self-circulating. Now a circulator is always used to pump heat through the system.

A central heating system is a closed system with either an expansion tank or open expansion vessel. A buffer tank can also be installed in the system.

A wide range of fuel types are used in central heating. Coal, coke, wood, oil, gas, wood chips and wood pellets have all proven adequate fuel sources in central heating boilers.

Components of Central Heating Systems

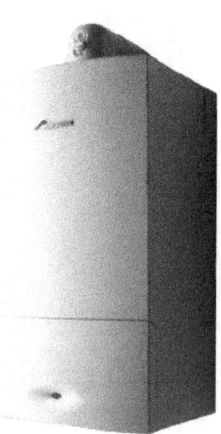

The Boiler

The boiler is the main component of a central heating system. They come in many sizes - delivering various amounts of heat energy, fuel types and energy ratings.

The Boiler Size

The amount of heat energy (measured in kilowatts or kW) a boiler is required to deliver through the home is based on a series of calculations carried out by an experienced and qualified plumber. They will be determined by the size of property, the building construction, building materials and how the boiler will be used.

Type of Fuel

There are a wide range of boilers that burn a range of fuel types. Below is a list of the most common types:

- Natural Gas - Burns methane from the gas mains in most towns and cities.
- LPG - Burns liquid petroleum gas, normally propane or butane.
- Oil type C2 - Burns kerosene which is the same as jet fuel.
- Oil type D - Burns 'gas oil' used mainly in oil Aga's.
- Solid mineral fuel - Burns coal or coke.
- Biomass fuel - Burns wood logs, pellets or chippings.
- Electric - Works like a kitchen kettle but on a much bigger scale.

SAP Rating

This is the Seasonal Efficiency Performance or the energy efficiency rating of a boiler and is listed as a band A to G. Type A is the best rating with 90% efficiency and type G is the worst with only 70% efficiency or below. SAP 2009 energy efficiency bands are as follows:

- Band A 90%

- Band B 86%-90%

- Band C 82%-86%

- Band D 78%-82%

- Band E 74%-78%

- Band F 70%-74%

- Band G Below 70%

The Building Regulations ensure that only the highest possible energy efficiency boilers are fitted and prevent low efficiency boilers being fitted.

Types of Boiler

- Conventional boiler - This boiler is the most basic type, it just burns fuel to make heat for central heating or hot water.

- System boiler - This boiler provides central heating only or heating and a store of hot water in a hot water cylinder (tank).

- Condensing -This boiler uses the heat in the gases given off when the fuel is burnt; this reusing of normally wasted heat makes some of the steam in the waste gases condense into water giving this boiler its name.

- Combination boiler - This boiler provides central heating and instant hot water.

Other Components of the Central Heating System

Heating Emitters - Radiators

The radiator is the most common way of heating your home. There are many types of radiator, many are made from copper or aluminium but most are made from steel.

A radiator works by transferring heat to the air in the room as it passes over the radiator panel. Warm air rises and pushes colder air back down and over the radiator surface again.

Heating Emitters - Underfloor Heating

Underfloor heating is a set of plastic pipes that are often run under a solid concrete floor surface and use the floor itself to heat the room, by radiating heat upwards.

This type of heating will generally, only be fitted during a new build or extension or conservatory added to an existing property.

Consideration will have to be given to the floor covering, tiles and wooden floors are ideal, however, deep pile carpets may have an adverse effect on the heating performance, acting as an insulator.

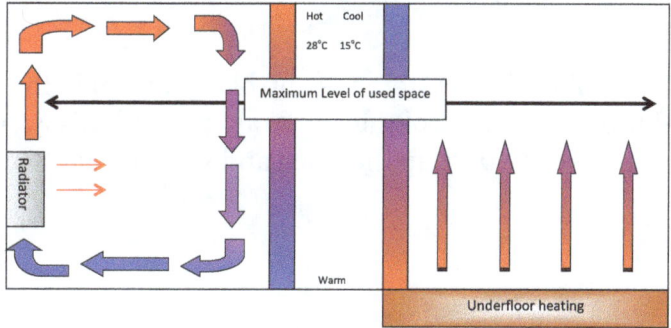

Heating Emitters - Fan Heaters

These work by using heated water passing through tubes that have a fan blowing air passed them to heat the air up.

Pipework

Pipes for central heating systems can be either copper or plastic, and come in many sizes from 8mm to 35mm in diameter. In domestic central heating systems the most common pipe sizes are 28mm, 22mm and 15mm.

The type of pipework used in a system is dependent upon a number of factors. An APHC member can provide advice and guidance on the best pipework material for an installation.

Expansion Vessel

This is used in a sealed central heating system to control expansion in the pipes and radiators, because as water is heated it gets bigger in volume by about 4%, and this water has to go somewhere. The expansion vessel is designed to take up the slack and stop the parts of the system bursting.

Expansion Relief Valve

This valve is also part of the 'sealed' heating system. It is designed to operate if there is a problem with the expansion vessel or the system is over pressurised. This operates to remove pressure from the system.

Header Tank

Plumbers call this little water tank (found in a loft or a high place) a 'feed and expansion' tank. Its job is to top up your central heating system with water, and if your system overheats it provides somewhere for the hot water to go. It is part of an 'open' heating system.

Pump

The pump is a part of your central heating system, without it the water heated from the boiler wouldn't go anywhere. This part can sometimes be located inside the boiler or can be found in your airing cupboard.

Motorised Valve

This component is responsible for choosing where the water from the boiler goes. The valves have a motor attached to the top so they are able to control the flow of heating water to either the central heating or hot water system. They come in two main types, 2 port or 3 port valves.

Central Heating Controls

Controls are the brain of the central heating system, with sensors and valves, the controls come together to make the system work.

Boiler Thermostat

A boiler will usually have a dial on it, marked in numbers or from Min to Max. This sets the temperature of the water that will be pumped from the boiler through the radiators to heat your home.

Programmer/Timer

The programmer/timer controls the flow of hot water to the radiators or hot water cylinder (if fitted), and it also decides to turn the system on or off depending on the temperature in the rooms and the time of day.

Room Thermostat

Individual room thermostats in your house measure the temperature of the air in the room, if it is cold it will tell the central heating to turn on, when it gets too warm it tells the central heating to turn off. Room thermostats need a free flow of air to sense the temperature, so they must not be blocked by curtains or furniture, or put near heat sources.

Thermostatic Radiator Valves (TRV)

These work in a similar way to the room thermostat, the TRV senses air temperature around them and turns the radiator on or off and regulates the flow of water through the radiator they are fitted to. They do not control the boiler directly.

Geothermal Heat Pump

Geothermal heat pumps refer to systems that use the ground, groundwater, or surface water as a heat source or sink. Specific to their configuration, these systems are referred to as ground-coupled heat pumps, groundwater heat pumps, and surface water heat pumps, respectively.

The geothermal heat pump is packaged in a single cabinet, and includes the compressor, loop-to-refrigerant heat exchanger, and controls. Systems that distribute heat using ducted air also contain the air handler, duct fan, filter, refrigerant-to-air heat exchanger, and condensate removal system for air conditioning. For home installations, the geothermal heat pump cabinet is usually located in a basement, attic, or closet. In commercial installations, it may be hung above a suspended ceiling or installed as a self-contained console.

Distribution Subsystem

Most residential geothermal systems use conventional ductwork to distribute hot or cold air and to provide humidity control. (A few systems use water-to-water heat pumps with one or more fan-coil units, baseboard radiators, or under-floor circulating pipes.) Properly sized, constructed, and sealed ducts are essential to maintain system efficiency. Ducts must be well insulated and, whenever possible, located inside of the building's thermal envelope (conditioned space).

Geothermal heating and cooling systems for large commercial buildings, such as schools and offices, often use a different arrangement. Multiple heat pumps (perhaps one for each classroom or office) are attached to the same earth connection by a loop inside the building. This way, each area of the building can be individually controlled. The heat pumps on the sunny side of the building may provide cooling while those on the shady side are providing heat. This arrangement is very economical, as heat is merely being transferred from one area of the building to another, with the earth connection serving as the heat source or heat sink only for the difference between the building's heating and cooling needs.

Water Heating

Many residential-sized systems installed today are equipped with desuperheaters to provide domestic hot water when the system is providing heat or air conditioning. The desuperheater is a small auxiliary heat exchanger at the compressor outlet. It transfers excess heat from the compressed gas to a water line that circulates water to the house's hot water tank. In summer, when the air conditioning runs frequently, a desuperheater may provide all the hot water needed by a household. It can provide four to eight gallons of hot water per ton of cooling capacity each hour it operates. A desuperheater provides less hot water during the winter, and none during the spring and fall when the system is not operating.

Because the heat pump is so much more efficient than other means of water heating, manufacturers are beginning to offer "triple function," "full condensing," or "full demand" systems that use a separate heat exchanger to meet all of a household hot water needs. These units cost-effectively provide hot water as quickly as any competing system.

The water heating system that is installed in the Finger Lakes Institute is an on demand system. This system provides hot water as soon as there is a demand for it. Using this type of system eliminates the need to heat stored water like a conventional hot water tank requires.

Types of Systems

Geothermal systems use the earth as a heat source and heat sink. A series of pipes, commonly called a "loop," carry a fluid used to connect the geothermal system's heat pump to the earth.

Closed and Open Loops

There are two basic types of loops: closed and open. Open loop systems are the simplest. Used successfully for decades, ground water is drawn from an aquifer through one well, passes through the heat pump's heat exchanger, and is discharged to the same aquifer through a second well at a distance from the first. Generally, two to three gallons per minute per ton of capacity are necessary for effective heat exchange. Since the temperature of ground water is nearly constant throughout the year, open loops are a popular option in areas where they are permitted. Open loop systems do have some associated challenges.

Some local ground water chemical conditions can lead to fouling the heat pump's heat exchanger. Such situations may require precautions to keep carbon dioxide and other gases in solution in the water. Other options include the use of cupronickel heat exchangers and heat exchangers that can be cleaned without introducing chemicals into the groundwater.

Increasing environmental concerns mean that local officials must be consulted to assure compliance with regulations concerning water use and acceptable water discharge methods. For example, discharge to a sanitary sewer system is rarely acceptable.

A closed loop system is being used for the Finger Lakes Institute. Closed loop systems are becoming the most common. When properly installed, they are economical, efficient, and reliable. Water (or a water and antifreeze solution) is circulated through a continuous buried pipe keeping. The closed loop system is environmentally friendly because water in the loop prevents contamination to the external environment.

The length of loop piping varies depending on ground temperature, thermal conductivity of the ground, soil moisture, and system design. (Some heat pumps work well with larger inlet temperature variations, which allows marginally smaller loops).

Horizontal Loops

Horizontal closed loop installations are generally most cost-effective for small installations, particularly for new construction where sufficient land area is available. These installations involve burying pipe in trenches dug with back-hoes or chain trenchers. Up to six pipes, usually in parallel connections, are buried in each trench, with minimum separations of a foot between pipes and ten to fifteen feet between trenches.

Vertical Loops

Vertical closed loops are preferred in many situations. For example, most large commercial buildings and schools use vertical loops because the land area required for horizontal loops would be prohibitive. Vertical loops are also used where the soil is too shallow for trenching. Vertical loops also minimize the disturbance to existing landscaping.

For vertical closed loop systems, a U-tube (more rarely, two U-tubes) is installed in a well drilled 100 to 400 feet deep. Because conditions in the ground may vary greatly, loop lengths can range from 130 to 300 feet per ton of heat exchange. Multiple drill holes are required for most installations, where the pipes are generally joined in parallel or series-parallel configurations.

A vertical loop well field, being used for the Finger Lakes Institute, consists of 20 wells, drilled to a depth of 100'. There are 5 (clusters) of 4 wells spaced approximately 12 feet on center, The depth and number of wells was determined by the estimated heat and cooling load required maintain a comfortable environment for the occupants.

Slinky Loops

Increasingly, "Slinky" coils overlapping coils of polyethylene pipe are used to increase the heat exchange per foot of trench, but require more pipe per ton of capacity. Two-pipe systems may require 200 to 300 feet of trench per ton of nominal heat exchange capacity. The trench length decreases as the number of pipes in the trench increases or as Slinky coil overlap increases. (Illustration below shows a slinky coil in a pond).

Pond Loops

Pond closed loops are a special kind of closed loop system. Where there is a pond or stream that is deep enough and with enough flow, closed loop coils can be placed on the pond bottom. Fluid is pumped just as for a conventional closed loop ground system where conditions are suitable, the economics are very attractive, and no aquatic system impacts.

Application

Geothermal heat pump systems allow for design flexibility and can be installed in both new and retrofit situations. Because the hardware requires less space than that needed by conventional heating, ventilating, and air-conditioning systems, the equipment rooms can be greatly scaled down in size, freeing space for productive use. Geothermal heat pump systems also provide excellent "zone" space conditioning, allowing different parts of the home to be heated or cooled to different temperatures.

Operation and Maintenance

Because geothermal heat pump systems have relatively few moving parts, and because those parts are sheltered inside a building, they are durable and highly reliable. The underground piping often carries warranties of 25 to 50 years, and the heat pumps often last 20 years or more. Since they usually have no outdoor compressors, they are not susceptible to vandalism. The components in the living space are easily accessible, which increases the convenience factor and helps ensure that the upkeep is done on a timely basis.

Special Considerations

Special considerations for geothermal heat pump systems include relevant codes and standards.

Relevant Codes and Standards

Design standards for geothermal direct-use systems typically involve two components:

1. Below-ground installation such as drilling wells, casing, and pumps.

2. Above-ground installations such as pipelines, pumps, valves, heat exchangers, in-building heat convectors, refrigeration equipment, and low temperature components such as heat pumps.

The below-ground equipment standards are usually specified for high temperatures (above 100° C) resources by state and country regulations and standards that would require special values, such as blow-out preventers and drilling muds. These are usually regulated and inspected by departments of geology and mineral industries or local level organizations. Low temperature resources (below 100° C) are usually regulated as standard water wells under the supervision of water resources departments or similar agencies.

Above-ground installation equipment standards are generally not regulated by geothermal requirements, but as standard off-the-shelf equipment.

Direct exchange geothermal heat pump is one in which the refrigerant circulates through a copper pipe placed directly in the ground. This eliminates the need for a heat exchanger between the refrigerant loop and the water loop, as well as eliminating the water pump. These simpler systems are able to reach higher efficiencies while also requiring a shorter and smaller pipe to be placed in the ground, reducing installation cost. DX systems are a relatively newer technology than water-source. DX systems, like water-source systems, can also be used to heat water in the house for use in radiant heating applications and for domestic hot water, as well as for cooling applications. Though corrosion or cracking of the copper loop has sometimes been a concern, these can be eliminated through proper installation. Since copper is a naturally-occurring metal that survives in the ground for thousands of years in most soil conditions, the copper loops usually have a very long lifetime.

Desuperheater

A desuperheater makes it possible for DX systems to provide almost free water heating throughout the summer. Instead of sending all the waste heat outdoors some can be captured and channeled to a water heater. Once the desuperheater has been installed (for about $400) the only cost of using it is the energy needed to circulate the water.

Advantages of DX Heat Pumps

DX systems offer a number of advantages beyond simply reducing energy consumption. These advantages fall into the following categories:

- Longer life and less maintenance

- Greater comfort.

Longer Life and Less Maintenance

Properly installed ground-source heat pumps cause less wear on the compressor than air-to-air heat pumps. This is because the temperature underground is more uniform than above ground so there is less stress on the system.

A DX system may be located entirely indoors and underground. This protects the equipment from the elements and temperature extremes while eliminating the need for costly defrost cycles, or outdoor fans. The result is fewer maintenance problems and longer equipment life than air-to-air systems. Heat Pump systems require less maintenance than oil or gas systems. DX systems do not emit combustion gases so therefore do not need to be vented. As a result DX heated homes are easier to build and can be built more energy efficiently.

Greater Comfort

Conventional forced air heating systems provide heat in short blasts which can dry out a house, making it uncomfortable and possibly even damaging it. DX systems, on the other hand provide a cooler, more prolonged flow of air.

DX systems can dehumidify a home better than standard central air conditioners, providing a higher level of comfort.

Air Source Heat Pumps

An air-source heat pump can provide efficient heating and cooling for your home. When properly installed, an air-source heat pump can deliver one-and-a-half to three times more heat energy to a home than the electrical energy it consumes. This is possible because a heat pump moves heat rather than converting it from a fuel like combustion heating systems do.

Air-source heat pumps have been used for many years in nearly all parts of the United States, but until recently they have not been used in areas that experienced extended periods of subfreezing temperatures. However, in recent years, air-source heat pump technology has advanced so that it now offers a legitimate space heating alternative in colder regions.

For example, when entire units are replaced in the Northeast and Mid-Atlantic regions, the Northeast Energy Efficiency Partnerships found that the annual savings when using an air-source heat pump are around 3,000 kWh (or $459) when compared to electric resistance heaters, and 6,200 kWh (or $948) when compared to oil systems. When displacing oil (i.e., the oil system remains, but operates less frequently), the average annual savings are near 3,000 kWh (or about $300).

Types of Air-Source Heat Pumps

The different types of air source heat pumps are described below.

Ductless vs. Ducted

Ductless applications require minimal construction as only a three-inch hole through the wall is required to connect the outdoor condenser and the indoor heads. Ductless systems are often installed in additions.

Ducted systems simply use ductwork. If your home already has a ventilation system or the home will be a new construction, you might consider this system.

Short-run ducted is traditional large ductwork that only runs through a small section of the house. Short-run ducted is often complemented by other ductless units for the remainder of the house.

Split vs. Packaged

Most heat pumps are split-systems—that is, they have one coil inside and one outside. Supply and return ducts connect to the indoor central fan.

Packaged systems usually have both coils and the fan outdoors. Heated or cooled air is delivered to the interior from ductwork that passes through a wall or roof.

Multi-zone vs. Single-zone

Single-zone systems are designed for a single room with one outdoor condenser matched to one indoor head.

Multi-zone installations can have two or more indoor heads connected to one outdoor condenser.

Multi-zone indoor heads vary by size and style and each creates its own "zone" of comfort, allowing you to heat or cool individual rooms, hallways, and open spaces. This distinction may also be referred to as "multi-head vs. single-head" and "multi-port vs. single-port."

Working of Air Source Heat Pumps

A heat pump's refrigeration system consists of a compressor and two coils made of copper tubing (one indoors and one outside), which are surrounded by aluminum fins to aid heat transfer. In heating mode, liquid refrigerant in the outside coils extracts heat from the air and evaporates into a gas. The indoor coils release heat from the refrigerant as it condenses back into a liquid. A reversing valve, near the compressor, can change the direction of the refrigerant flow for cooling as well as for defrosting the outdoor coils in winter.

The efficiency and performance of today's air-source heat pumps is a result of technical advances such as the following:

- Thermostatic expansion valves for more precise control of the refrigerant flow to the indoor coil.

- Variable speed blowers, which are more efficient and can compensate for some of the adverse effects of restricted ducts, dirty filters, and dirty coils.

- Improved coil design.

- Improved electric motor and two-speed compressor designs.

- Copper tubing, grooved inside to increase surface area.

Selecting a Heat Pump

Heating efficiency for air-source electric heat pumps is indicated by the heating season performance factor (HSPF), which is the total space heating required during the heating season, expressed in Btu, divided by the total electrical energy consumed by the heat pump system during the same season, expressed in watt-hours.

Cooling efficiency is indicated by the seasonal energy efficiency ratio (SEER), which is the total heat removed from the conditioned space during the annual cooling season, expressed in Btu, divided by the total electrical energy consumed by the heat pump during the same season, expressed in watt-hours.

The HSPF rates both the efficiency of the compressor and the electric-resistance elements.

The SEER rates a heat pump's cooling efficiency. In general, the higher the SEER, the higher the cost. However, the energy savings can return the higher initial investment several times during the heat pump's life. A new central heat pump replacing a vintage unit will use much less energy, cutting air-conditioning costs substantially.

These are some other factors to consider when choosing and installing air-source heat pumps:

- Select a heat pump with a demand-defrost control. This will minimize the defrost cycles, thereby reducing supplementary and heat pump energy use.

- Fans and compressors make noise. Locate the outdoor unit away from windows and adjacent buildings, and select a heat pump with an outdoor sound rating of 7.6 bels or lower. You can also reduce this noise by mounting the unit on a noise-absorbing base.

- The location of the outdoor unit may affect its efficiency. Outdoor units should be protected from high winds, which can cause defrosting problems. You can strategically place a bush or a fence upwind of the coils to block the unit from high winds.

Measures of Efficiency and Benefits of Air Source Heat Pumps

Air source heat pumps performances are measured through a Coefficient of Performance (COP) that can have different values that mean how many units of heat are produced using one unit of energy.

There are main advantages related to this technology both on environmental and economic sides. First of all, air source heat pumps don't have an environmental impact as significant as the heat they use for the process is extracted either by air, water or ground and it is continuously regenerated although they still make use of electricity in the process. On the financial side it can allow expenses savings with respect to electricity-powered heatings, it is supported by the State through the Renewable Heat Incentive and householders can reduce carbon emissions by cutting on harmful fuels. Furthermore, this technology does not need frequent maintenance but it usually works smoothly after the installation and it is cheaper to install than ground source pumps as it does not need any kind of excavation site. However, it could be less efficient than the ground pump and its performance can be negatively affected by low temperatures and it usually needs a longer time and bigger surfaces to heat the interiors.

Advanced Technologies: Reverse Cycle Chillers

One of the more notable innovations in air-source heat pumps is called a reverse cycle chiller (RCC). It offers the advantages of allowing the homeowner to choose from a wide variety of heating and cooling distribution systems, from radiant floor systems to forced air systems with multiple zones. It also offers the potential for lower winter electric bills and hotter air out of the supply vents for greater comfort.

An RCC is especially economical for all-electric homes or in areas where natural gas is not available. Depending on other fuel rates, it may even be the least expensive heating option among the remaining heating fuel choices.

The system consists of a standard single speed, air-source heat pump, sized to the heating load rather than the usual smaller summer cooling load. The heat pump is connected to a large, heavily insulated tank of water that the heat pump heats or cools, depending on the season of the year. Most systems will use a fan coil with ducts, using the stored water to heat or cool the air and distribute it to the house.

The RCC system allows the heat pump to operate at peak efficiency even at low temperatures. This provides greater comfort and economy without the need for electric resistance auxiliary heating coils.

The RCC can also be equipped with a refrigeration heat reclaimer, which is similar to the common desuperheater coil found on the high-end heat pumps and air conditioners. The combined system costs about 25% more than a standard heat pump of similar size, and the simple payback on the additional cost in areas where natural gas is not available is about 2 to 3 years.

Ground-coupled Heat Exchanger

A Ground Coupled Heat Exchanger Air-conditioning System (GCHE) is a modern type of space heating & cooling system which exchanges heat with the ground or earth rather than the ambient air. The temperature of the ambient air fluctuates throughout the year but the temperature of ground at a certain depth remains constant throughout the year all around the globe. The ground coupled heat exchanger as exchanges heat with nearly constant temperature throughout the globe heat sink (ground) rather than exchanging it with fluctuating temperature heat-sink(ambient air) consumes nearly same amount of electricity throughout the year if the cooling load is kept constant as the heat sink is at nearly constant temperature, whereas in case of ambient air as heat-sink the electricity consumption increases/decreases as the temperature of the sink fluctuates i.e. it increases in the summer due to which more amount of work and hence energy is required to pump heat from room to be cooled to a already hotter heat-sink whereas the temperature of the heat-sink in winter is low so energy consumption is comparatively low. Generally, cooling load is considered in summer season, hence comparatively ground or earth as a heat-sink is an economical way than conventional system having ambient air as heat-sink when electricity consumption is considered.

A GCHE system eliminates the requirement of water required for cooling towers thus saving the water lost through evaporation in cooling waters.

Principle Of Operation

The GCHE system works on the phenomena of earth having constant temperature throughout the year around the globe below a certain depth beneath the earth surface. Usually, it is 15° C-20° C but may vary according to certain geological and geographical conditions .The heat transfer rate between the GCHE and earth depends upon the conductivity of soil, thermal inertia of soil along with many other parameters like water holding capacity of the soil, and the depth.

The coils of GCHE are made up of High Density Polyurethane (HDPE) pipes.

The cold fluid i.e. water from the GCHE enters the indoor unit of the air-conditioning system extracts heat from the room or space to be cooled and returns to the GCHE where it exchanges heat with the earth at constant temperature and gets cooled again. The point at which the warm liquid enters the GCHE is hotter than the intermediate point between the GCHE and the point at which the liquid exits the GCHE.

The vertical configuration is more effective when there is a limited space available, the horizontal loop is used when more space is available and soil conductivity and water holding capacity is significant, whereas the pond loop configuration gives the added advantage of extracting heat faster from the heated liquid as the water flows over the GCHE coils.

The GCHE system if coupled with Earth Air Heat Exchanger (EAHE) system increases the efficiency of GCHE system. The EAHE system passes air from the coils buried at a certain depth so that the air is pre-cooled before letting it into the room to be cooled.

Space Heater

Space heater is a usually portable appliance for heating a relatively small area.

When the temperature falls, sometimes the best option for heating up a room is with a space heater. No need to bust out the winter coat or grab a thick blanket; a space heater will do the trick nicely.

There are many options out there for people looking to keep warm. Lots of heating systems boast tons of new features and capabilities, but if you're looking for a simple way to create heat, natural gas or propane space heaters are a great way to do it.

Powered by natural gas or liquid propane, these space heaters can be used in case of emergency or a power outage since they don't require electricity to run. They're perfect for your home, especially when heating up a small area.

Space heaters for homes work by first heating up the surrounding air, then moving that air all around the room. This is an efficient way to create warmth while also saving money in comparison to central heating systems.

Vented And Unvented Combustion Space Heaters

Space heaters are classified as vented and unvented or "vent-free." Unvented combustion units are not recommended for use inside your home, because they introduce unwanted combustion products into the living space – including nitrogen oxides, carbon monoxide, and water vapor – and deplete air in the space. Most states have banned unvented kerosene heaters for use in the home and at least five have banned the use of unvented natural gas heaters.

Vented units are designed to be permanently located next to an outside wall, so that the flue gas vent can be installed through a ceiling or directly through the wall to the outside. Look for sealed combustion or "100% outdoor air" units, which have a duct to bring outside air into the combustion chamber. Sealed combustion heaters are much safer to operate than other types of space heaters, and operate more efficiently because they do not draw in the heated air from the room and exhaust it to the outdoors. They are also less likely to backdraft and adversely affect indoor air quality.

Electric Space Heaters

Electric space heaters are generally more expensive to operate than combustion space heaters, but they are the only unvented space heaters that are safe to operate inside your home. Although electric space heaters avoid indoor air quality concerns, they still pose burn and fire hazards and should be used with caution.

For convection (non-radiant) space heaters, the best types incorporate a heat transfer liquid, such as oil, that is heated by the electric element. The heat transfer fluid provides some heat storage, allowing the heater to cycle less and to provide a more constant heat source.

Types of Portable Electric Space Heaters

Portable electric space heaters come in two main types: radiant and convection.

Radiant Heaters

Radiant heaters produce infrared radiation that warms bodies and objects rather than the air. Because they direct heat towards a specific area rather than circulate it through the air (they act essentially as "personal" space heaters), they are good for small rooms or spot heating right where one or two people are working in a basement or garage.

- Metal-rod radiant heaters use a heat element placed in front of a shiny reflector that radiates heat.

- Quartz radiant heaters use electric elements packed inside a quartz glass tube that radiates heat.

Convection Heaters

Convection heaters provide warmth by blowing or pulling air over a heated surface and circulating it through the air. They are the best choice for quickly warming larger spaces and keeping them warm over a longer period.

- Electric-element convection heaters move air over heated wire or coil elements inside the heater. These heaters can heat up a space quickly, but the heat will quickly disperse into the air and the heater will need to cycle on regularly to maintain temperature levels.

- Liquid-filled convection heaters use an electric element to warm a liquid, usually either water or oil. This liquid in turn heats a radiator to create natural air currents. These heaters take longer to heat up than electric-element heaters, but will hold their heat longer, minimizing temperature fluctuations.

- Ceramic convection heaters use a fan to draw air through a ceramic element and blow it through a room. These are the safest electric space heaters because the electric element never gets hot enough to start a fire.

Underfloor Heating

Underfloor or radiant heating is an essential part of the central heating system, which generates heat through a system of interconnected pipes or electric wires, which are mounted beneath or within the floor coating. The underfloor heating occurs once electricity is passed through the wires, or hot fluids (which usually is a mix of water and antifreeze) are pumped into the tubes beneath the floor. Subsequently, the produced heat is transferred to the floor's upper layer, which later sustains an upward heating effect which gradually warms the room space from bottom to top, creating a pleasing feeling of warmth.

An underfloor heating system comes in two variants: electric and water underfloor heating. While the electric one is cheaper to install but more expensive in exploitation, the water-based heating performs better in the long run but requires some floor renovation works that do not come cheap. Choosing between these two comes down to your budget allowance and to the heat source that's being used in your home's central heating system.

Electric Underfloor Heating

An electric radiant heating system represents a web of electric non-corrosive cables arranged in a spiral type structure, that are connected to a power supply and a thermostat for the purpose of controlling the heat output. When installing an electric underfloor heating system you'll have to consider the floor's type, how well it is insulated and the size of the room your are looking to heat. A system like this is good for rooms with stone, tiled or even carpeted floors, since it does not require too much ground space and can be placed close to the surface, making it easier for those home-owners that do not foresee a major floor refurbishment. Although the costs of operating an electric underfloor heating can be somewhat higher compared to the ones of a water based system (due to its dependency on electricity as means of heating the metal spirals), it bears low installation costs and it is quite reliable during exploitation.

Electric Floor Heating Installation

Depending on the preference of the home owner and the application, these systems can be installed using loose cables or mats. There are several different systems that suit a range of design requirements.

Electric Heating Cables in Screed Bed

In screed heating consists of heating cables fixed on top of the floor structure with a light steel mesh or fixing strips. The cables are embedded in a sand and cement bed which becomes heated, warming the floor covering. In screed heating is considered a direct acting system, although it does offer some heat storage.

This system is the best option for rooms that have some flexibility in the height of the floor. It is typically embedded in 20-30 mm of screed, usually at the time of construction.

Because of this, these systems can raise the floor of an existing room by 30 mm in addition to the floor tile. The benefit of this system lies in its ability to hold the heat for a much longer period when compared to an adhesive system.

These systems tend to take up to 30-45 minutes to heat up and give off heat for the same period once they have been turned off. Screed bed systems work well in areas that are irregularly shaped, although it is recommended that they are installed by a trained professional.

Electric Heating Cables in Concrete Floor Slab

In slab heating consists of heating cable that is typically fixed to the top layer of the slab reinforcement, prior to the slab being poured. The heating cables heat up the slab, in turn warming the floor covering above. Cables are laid on the top mesh in about 30-40 mm of concrete and run during off-peak hours to cut energy costs and maximise the efficiency of the system.

As a storage system, in peak hours it runs on energy that has been accumulated during off-peak hours. This system is usually used in new construction before the slab is laid.

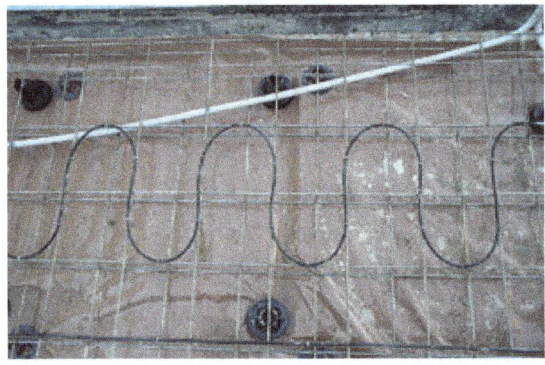

Electric Heating Cables in Tile Adhesive

Under tile heating consists of a thin heating cable that is pre-laid on a self-adhesive fibreglass mesh. It sits in the glue or levelling screed which then becomes heated and warms the floor covering.

Cables can be placed in a tile adhesive of at least 5–15 mm depending on the design requirements of the room and the personal preference of the home owner. This is an ideal option for retrofitting and new construction projects.

This "on-demand" system is quick to respond to changes in the thermostatic reading. Located immediately under the heated surface, it tends to be the fastest floor heating system on the market.

The installation costs for these systems can be on the high side, but running costs have the potential to be quite low.

Water Underfloor Heating

If an electric underfloor heating uses a network of wires to generate heat, then the water underfloor heating uses a series of small pipes installed underneath the floor for the same purpose.

Hydronic underfloor heating is an energy efficient and sustainable solution for heating your home. It works well under several different floor surfaces including marble, stone, slate, carpet, timber and all varieties of tiles.

Hydronic underfloor heating is a central heating system and is designed to run continuously during the (winter) heating season. It is not designed for occasional "demand" heating. Where greater heating flexibility is required, particularly for small areas such as bathrooms, en-suites, laundries etc., electric floor heating is recommended.

Hydronic floor heating has become an attractive and regularly specified upgrade for a wide range of applications, especially when paired with a high efficiency condensing natural gas boiler. These systems are quite sophisticated in design and consist of warm water that circulates through an intricate network of polyethylene oxygen-resistant piping under the floor surface. This ensures a gentle, warm heat that is evenly distributed throughout the entire living space.

Maintenance

These systems require a certain level of maintenance over the years. Boilers require regular servicing and may need to be replaced after 10 years of use. Good quality pipes generally last a very long time.

The overall system is known to exceed thirty to forty years in lifespan. On top of the fairly low running cost, this can give owners a high return on investment when measured against other heating solutions on the market.

Cost

Hydronic floor heating can be very economical to run, however, the upfront costs tend to be higher than electric systems, because of the sophistication and intricacy of the design, as well as the skilled labour that is required to perform a proper installation.

A significant part of a hydronic floor heating costs is fixed, meaning it remains similar whatever the size of the area to heat. A gas boiler is an example of a fixed cost that hardly changes when the size of the surface to heat increases. For this reason, the bigger the surface, the cheaper the rate per m2 is. This is an important difference with Electric Floor heating, where the rate per m2 remains the same as the surface to heat increases. This is the reason why hydronic floor heating (HFH) is a more and more interesting proposition than Electric Floor Heating (EFH) as the area to heat increases, and we would recommend to go for HFH over EFH for surface over 60m².

Thermostats

Another factor in the efficiency of your floor heating system is a high quality thermostat. This will allow you to heat specific rooms independently and program them to meet the needs of your lifestyle.

Because hydronic floor heating is a central heating system, we recommend non-programmable thermostats for your installation. These thermostats can be linked to your home automation systems.

A good quality thermostat will give greater control over temperature, zoning and timing, allowing you to make the most out of your floor heating system.

Installation

Installation of a water-based system can be complex due to the intricate design and workings of the system. It is important to seek out a licensed professional to do the job.

Hydronic floor heating systems are normally only used in new projects because it is easier to install into the slab. In some circumstances these systems can be used in retrofit projects, but only in a screed bed that is at least 50 mm in depth (ideally 80mm). As this is a significant height, some customers prefer to go for electric floor heating for renovations.

In hydronic floor heating systems, water is warmed up to 50 degrees Celsius and circulates at a safe, low pressure through a network of pipes, valves, manifolds and switches, all of which must work together to heat the zoned areas.

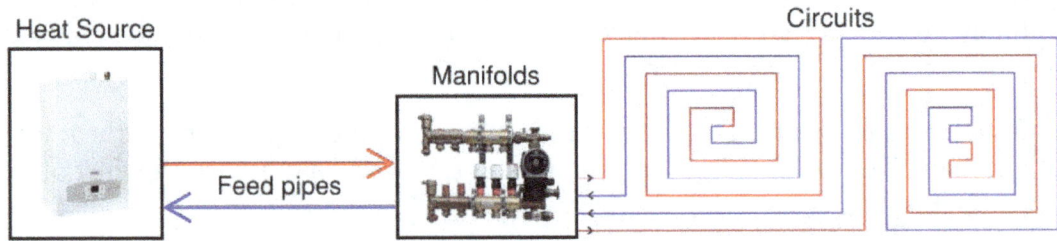

It is a closed system, which means that the water that is introduced the day the system is commissionned is most likely to be the same for the rest of the life of the hydronic system. For this reason, it is crucial to select the best quality for the manifolds, pipes, boilers and anything that will be in contact with the water, because any inpurity will worsen over time.

Heating Film

The heating film is the high quality heating product used mainly for: underfloor heating, wall heating and ceiling heating, but the functionality of the product along with the unlimited human imagination creates the new usage of the heating film. The non standard usage of the heating film is the terrarium and camp site heating or used as a main heating element in radiators, platforms or heating pillows. The heating film is a hit in the recent years in the branch of heating in churches, where the local electric underfloor heating is a guarantee of the highest comfort and lowest exploitation cost.

Today, there is no better, more advanced heating technology than a heating film. The high quality product will ensure you with the maximal thermal comfort in the whole house. The thin, but extraordinarily efficient heating film will let you heat effectively almost every room. As the whole structure is installed directly under the panels, the technology is almost invisible. The full automation lets regulating the heat level according to our expectations. The electric underfloor heating with the help of the infrared heating film has a significant influence on the well-being of people in the room – it not only provides the maximal comfort in use but it has also an impact on the immune system.

Raw Materials of Heat Film

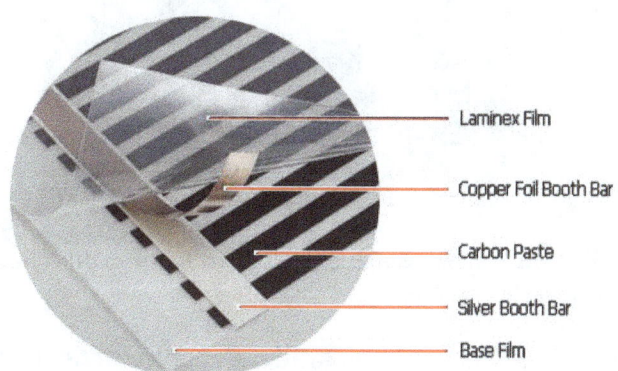

Laminex Film

Copper Foil Booth Bar

Carbon Paste

Silver Booth Bar

Base Film

Insulation/Flame-Retardant PET Film

PET is abbreviation for polyethylene terephthalate, part of the thermoplastics. It is light, has no taste or smell and it is widely used in household items, toys, electric insulators, radio and TV case, and packing materials. For Hot-Film, PET meets the VTM-2 standard requirement of insulating/flame-retardant UL.

It has a color of milk and is used for electronic/electric materials.

Carbon Paste

Is a fine black powder with carbon being the main element. It has a particle size is 1 ~500mi and is used to make a black paint(pigment) and rubber and printing ink.

For the heating element of carbon, the quantity of heating is controllable by adjusting its resistance value based on the mixture ratio of carbon and graphite.

Silver Paste

Silver booth bar is used for increased conductivity and minimize electric sparks caused by a direct contact between copper foil and carbon heating section. An ideal resistance value should be 1 Q or less.

Laminex Film

Laminex film which is used in heating film manufacturing has high adhesive strength, high heat-resisting and easily processing property.

Especially, its adhesive substance is using EVA(for under-floor) or EEA(for sauna).

Benefits of Heating Film

Easy-to-install and Economical Heating System

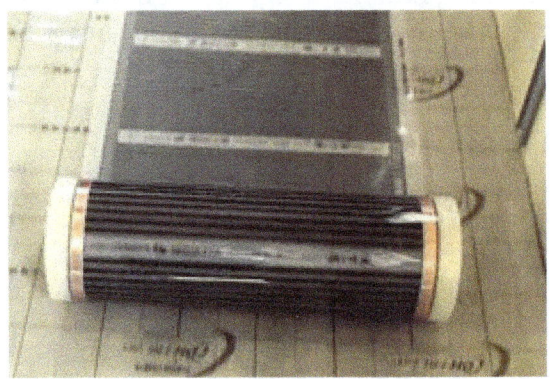

Capable of reducing installation expenses due to an *easy-to-construction/installation* process meaning reducing construction times.

Capable of having a fast heating effect due to a parallel structure of heating element and use of electricity. Capable of heating partially and saving unnecessary heating expenses (Possible to adopt a central control system) Capable of having a more capacious living space by not needing to have a separate boiler space.

Eco-friendly and Healthy Heating System

Heating element is made of carbon, such as charcoal (activated carbon), emitting less electronic waves. Far-infrared ray and anion suppresses Sick House Syndrome, odor, and growth of germs.

It does not use flames, so it generates less noise, dust, and carbon monoxide. It is perfect for areas with children, senior citizens and patients.

Maintenance Cost-saving Heating System

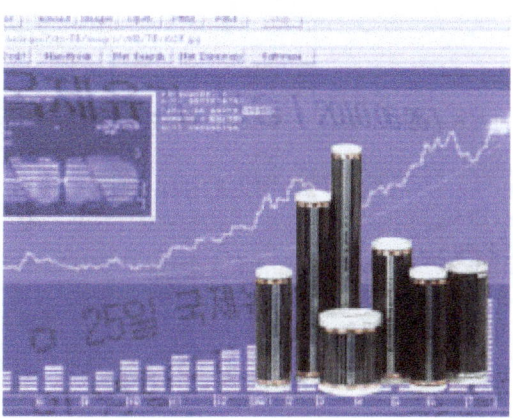

30% cost efficient than electric heating and over 50% cost efficient than oil boiler heating. Simple structure and design meaning fewer product failures and maintenance costs. For house remodeling, it is easy to remove and relocate the system.

Heat-film Technical Features

- Capable of keeping an electric resistance value stable due to printing technology using carbon particle of regularity and high density.

- Technology and development of high-tech printing equipment that enables a uniform printing side.

- Development of special laminating equipment for heating film production.

- Use of the world-class insulting and flame-retardant film.

- Technology of preventing sparks from carbon printing side due to design of special silver booth bar.

- Design of preventing the occurrence of sparks by adjusting sectional area of carbon printing side.

References

- Geothermal-heat-pumps: wbdg.org, Retrieved 15 July 2018
- Types-of-geothermal-heat-pumps: solairehomecomfort.com, Retrieved 22 June 2018
- How-do-air-source-heat-pumps-work: greenmatch.co.uk, Retrieved 29 May 2018
- Ground-Coupled-Heat-Exchanger-Air-Conditioning-System-A-Study: ijser.org, Retrieved 30 March 2018
- Resource-space-heater-safety: reinhardthomeheating.com, Retrieved 13 April 2018
- Underfloor-heating: greenmatch.co.uk, Retrieved 23 April 2018
- Electric-underfloor-heating: devexsystems.com.au, Retrieved 08 May 2018
- Hydronic-underfloor-heating: devexsystems.com.au, Retrieved 18 March 2018

4

Air Conditioning

Air conditioning is the technology of removing moisture and heat from an interior space for the comfort of the occupants. This chapter has been carefully written to provide an easy understanding of the varied aspects of air conditioning, such as district cooling system, chilled beam system, evaporative cooler, deep water source cooling, etc.

Air-conditioning is that process used to create and maintain certain temperature, relative humidity and air purity conditions in indoor spaces. This process is typically applied to maintain a level of personal comfort.

It's also used in industrial applications to ensure correct operation of equipment or machinery that need to operate in specific environmental conditions or alternatively to be able to carry out certain industrial processes, such as welding, which produce considerable amounts of heat that needs to be disposed of in some manner.

An air-conditioning system must be effective regardless of outside climatic conditions and involves control over four fundamental variables: air temperature, humidity, movement and quality.

The distinction between industrial and personal comfort applications is not always clear cut. Industrial air-conditioning usually requires better precision as regards temperature and humidity control. Some application also demand a high degree of filtering and removal of contaminants.

Comfort air-conditioning on the other hand, as well as needing to satisfy personal temperature-humidity requirements, also involves other fields such as architectural design, weather forecasting, energy consumption and sound emissions to recreate the ideal conditions for human psychophysiological well-being.

The air conditioning unit uses chemicals that convert from gas to liquid and back again quickly. These chemicals transfer the heat from the air inside our property to the outside air.

The AC unit has three key parts. These are the compressor, the condenser, and the evaporator. AC unit's compressor and condenser are typically located in the outside part of the air conditioning system. Inside the house is where we will find the evaporator.

The cooling fluid reaches the compressor as a low-pressure gas. The compressor squeezes this gas/fluid, and the molecules in the liquid are packed closer together. The closer the compressor forces these molecules together, the higher the temperature and energy rise.

Working of Air Conditioner

This working fluid exits the compressor as a high-pressure, hot gas, and it moves to the condenser. The outside unit of an air conditioning system has metal fins all around the housing. These fins work like the radiator on a vehicle, and they help dissipate heat more quickly.

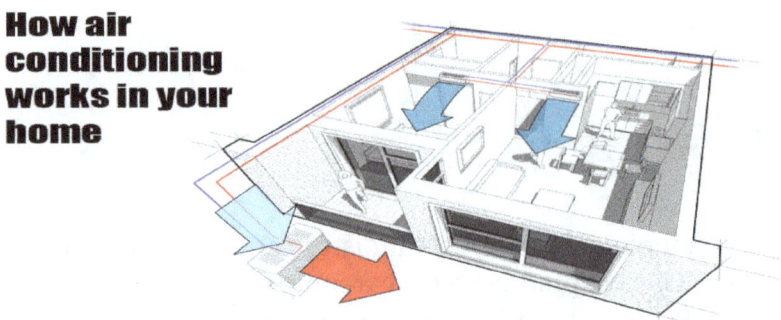

How air conditioning works in your home

This is how an air conditioner works to cool our home:

When the fluid leaves the condenser, it is much cooler. It's also changed from a gas to liquid because of the high pressure. The fluid makes its way into the evaporator through a minuscule, narrow hole and when the liquid reaches the other side of this passage, its pressure drops. When this happens, the fluid begins to evaporate to gas.

As this occurs, the heat is extracted from the surrounding air. This heat is required to separate the molecules of the liquid into a gas. The metal fins on the evaporator also help exchange thermal energy with the surrounding air.

When the refrigerant leaves the evaporator, it is once again a low-pressure, chilled gas. The process starts all over when it goes back to the compressor. There is a fan that's connected to the evaporator, and it circulates air around the inside of the property and across the fins of the evaporator.

The air conditioner sucks air into the ducts through a vent. This air is used to cool gas in the evaporator, and as the heat is removed from the air, it's cooled. Ducts then blow air back into the house.

This process continues until the inside air of our home or business reaches the desired temperature. When the thermostat senses that the interior temperature is at the desired level, it shuts the air conditioner off. When the room heats up again, the thermostat turns the air conditioner back on until the preferred ambient temperature is achieved again.

Humidity Control

Since humans perspire to provide natural cooling by the evaporation of perspiration from the skin, drier air (up to a point) improves the comfort provided. The comfort air conditioner is designed to create a 50% to 60% relative humidity in the occupied space.

Dehumidification and Cooling

Refrigeration air conditioning equipment usually reduces the absolute humidity of the air processed by the system. The relatively cold (below the dewpoint) evaporator coil condenses water vapor from the processed air, much like an ice-cold drink will condense water on the outside of a glass. Therefore, water vapor is removed from the cooled air and the relative humidity in the room is lowered. The water is usually sent to a drain or may simply drip onto the ground outdoors. The heat is rejected by the condenser which is located outside of room to be cooled.

Dehumidification Program

Most modern air-conditioning systems feature a dehumidification cycle during which the compressor runs while the fan is slowed as much as possible to reduce the evaporator temperature and therefore condense more water. When the temperature falls below a threshold, both the fan and compressor are shut off to mitigate further temperature drops; this prevents moisture on the evaporator from being blown back into the room. When the temperature rises again, the compressor restarts and the fan returns to low speed.

Occasionally, to thaw any ice produced, the fan runs with the compressor shut down; this function is less effective when ambient temperatures are low.

Inverter air conditioners use the inside coil temperature sensor to keep the evaporator as cold as possible. When the evaporator is too cold, the compressor is slowed or stopped with the indoor fan running.

Typical portable dehumidifier

A specialized air conditioner that is used only for dehumidifying is called a dehumidifier. It also uses a refrigeration cycle, but differs from a standard air conditioner in that both the evaporator and the condenser are placed in the same air path. A standard air conditioner transfers heat energy out of the room because its condenser coil releases heat outside. However, since all components of the dehumidifier are in the *same* room, no heat energy is removed. Instead, the electric power consumed by the dehumidifier remains in the room as heat, so the room is actually *heated*, just as by an electric heater that draws the same amount of power.

In addition, if water is condensed in the room, the amount of heat previously needed to evaporate that water also is re-released in the room (the latent heat of vaporization). The dehumidification process is the inverse of adding water to the room with an evaporative cooler, and instead releases heat. Therefore, an in-room dehumidifier always will warm the room and reduce the relative humidity indirectly, as well as reducing the humidity directly by condensing and removing water.

Inside the unit, the air passes over the evaporator coil first, and is cooled and dehumidified. The now dehumidified, cold air then passes over the condenser coil where it is warmed up again. Then the air is released back into the room. The unit produces warm, dehumidified air and can usually be placed freely in the environment (room) that is to be conditioned.

Dehumidifiers are commonly used in cold, damp climates to prevent mold growth indoors, especially in basements. They are also used to protect sensitive equipment from the adverse effects of excessive humidity in tropical countries.

Energy Transfer

In a thermodynamically closed system, any power dissipated into the system that is being maintained at a set temperature (which is a standard mode of operation for modern air conditioners) requires that the rate of energy removal by the air conditioner increase. This increase has the effect that, for each unit of energy input into the system (say to power a light bulb in the closed system), the air conditioner removes that energy. To do so, the air conditioner must increase its power consumption by the inverse of its "efficiency" (coefficient of performance) times the amount of power dissipated into the system. As an example, assume that inside the closed system a 100 W heating element is activated, and the air conditioner has a coefficient of performance of 200%. The air conditioner's power consumption will increase by 50 W to compensate for this, thus making the 100 W heating element cost a total of 150 W of power.

It is typical for air conditioners to operate at "efficiencies" of significantly greater than 100%. However, it may be noted that the input electrical energy is of higher thermodynamic quality (lower entropy) than the output thermal energy (heat energy).

Air conditioner equipment power in the U.S. is often described in terms of "tons of refrigeration", with each approximately equal to the cooling power of one short ton (2000 pounds or 907 kilograms) of ice melting in a 24-hour period. The value is defined as 12,000 BTU per hour, or 3517 watts. Residential central air systems are usually from 1 to 5 tons (3.5 to 18 kW) in capacity.

Seasonal Energy Efficiency Ratio

For residential homes, some countries set minimum requirements for energy efficiency. In the United States, the efficiency of air conditioners is often (but not always) rated by the *seasonal energy efficiency ratio (SEER)*. The higher the SEER rating, the more energy efficient is the air conditioner. The SEER rating is the BTU of cooling output during its normal annual usage divided by the total electric energy input in watt hours (W·h) during the same period.

> SEER = BTU ÷ (W·h)

this can also be rewritten as:

> SEER = (BTU/h) ÷ W, where "W" is the average electrical power in Watts, and (BTU/h) is the rated cooling power.

For example, a 5000 BTU/h air-conditioning unit, with a SEER of 10, would consume 5000/10 = 500 Watts of power on average.

annual operating time:

> 500 W × 1000 h = 500,000 W·h = 500 kWh

Assuming 1000 hours of operation during a typical cooling season (i.e., 8 hours per day for 125 days per year).

Another method that yields the same result, is to calculate the total annual cooling output:

> 5000 BTU/h × 1000 h = 5,000,000 BTU

Then, for a SEER of 10, the annual electrical energy usage would be:

> 5,000,000 BTU ÷ 10 = 500,000 W·h = 500 kWh

SEER is related to the coefficient of performance (COP) commonly used in thermodynamics and also to the Energy Efficiency Ratio (EER). The EER is the efficiency rating for the equipment at a particular pair of external and internal temperatures, while SEER is calculated over a whole range of external temperatures (i.e., the temperature distribution for the geographical location of the SEER test). SEER is unusual in that it is composed of an Imperial unit divided by an SI unit. The COP is a ratio with the same metric units of energy (joules) in both the numerator and denominator. They cancel out, leaving a dimensionless quantity. Formulas for the approximate conversion between SEER and EER or COP are available.

> SEER = EER ÷ 0.9
>
> SEER = COP × 3.792
>
> EER = COP × 3.413

From second equation above, a SEER of 13 is equivalent to a COP of 3.43, which means that 3.43 units of heat energy are pumped per unit of work energy.

The United States now requires that residential systems manufactured in 2006 have a minimum SEER rating of 13 (although window-box systems are exempt from this law, so their SEER is still around 10).

Installation Types

Window Unit and Packaged Terminal

How a window air conditioner works

Air conditioning window unit

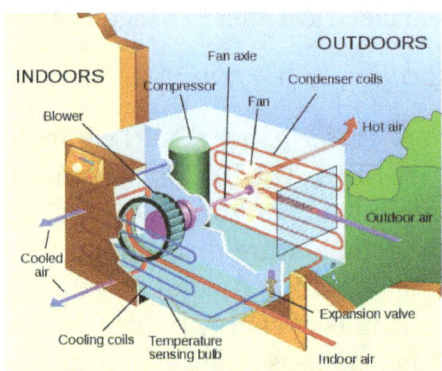

Parts of a window unit

Window unit air conditioners are installed in an open window. The interior air is cooled as a fan blows it over the evaporator. On the exterior the heat drawn from the interior is dissipated into the environment as a second fan blows outside air over the condenser. A large house or building may have several such units, allowing each room to be cooled separately.

In 1971, General Electric introduced a popular portable in-window air conditioner designed for convenience and portability.

Packaged terminal air conditioner (PTAC) systems are also known as wall-split air conditioning systems. They are ductless systems. PTACs, which are frequently used in hotels, have two separate units (terminal packages), the evaporative unit on the interior and the condensing unit on the exterior, with an opening passing through the wall and connecting them. This minimizes the interior system footprint and allows each room to be adjusted independently. PTAC systems may be adapted to provide heating in cold weather, either directly by using an electric strip, gas, or other heater, or by reversing the refrigerant flow to heat the interior and draw heat from the exterior air, converting the air conditioner into a heat pump. While room air conditioning provides maximum flexibility, when used to cool many rooms at a time it is generally more expensive than central air conditioning.

The first practical semi-portable air conditioning unit was invented by engineers at Chrysler Motors and offered for sale starting in 1935.

Split Systems

Split-system air conditioners come in two forms: mini-split and central systems. In both types, the inside-environment (evaporative) heat exchanger is separated by some distance from the outside-environment (condensing unit) heat exchanger.

Mini-split (Ductless) System

Outside part of a ductless split-type air conditioner

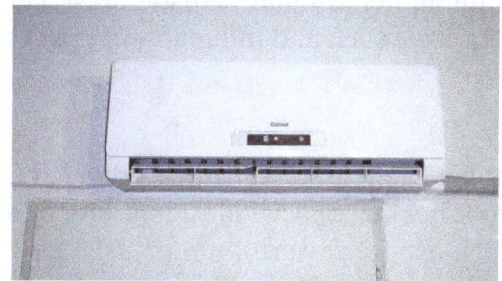

Indoor part of a ductless split-type air conditioner

A mini-split system typically supplies air conditioned and heated air to a single or a few rooms of a building. Multi-zone systems are a common application of ductless systems and allow up to 8 rooms (zones) to be conditioned from a single outdoor unit. Multi-zone systems typically offer a variety of indoor unit styles including wall-mounted, ceiling-mounted, ceiling recessed, and horizontal ducted. Mini-split systems typically produce 9,000 to 36,000 Btu (9,500–38,000 kJ) per hour of cooling. Multi-zone systems provide extended cooling and heating capacity up to 60,000 Btu's.

Advantages of the ductless system include smaller size and flexibility for zoning or heating and cooling individual rooms. The inside wall space required is significantly reduced. Also, the compressor and heat exchanger can be located farther away from the inside space, rather than merely on the other side of the same unit as in a PTAC or window air conditioner. Flexible exterior hoses lead from the outside unit to the interior one(s); these are often enclosed with metal to look like common drainpipes from the roof. In addition, ductless systems offer higher efficiency, reaching above 30 SEER.

The primary disadvantage of ductless air conditioners is their cost. Such systems cost about US $1,500 to US $2,000 per ton (12,000 BTU per hour) of cooling capacity. This is about 30% more than central systems (not including ductwork) and may cost more than twice as much as window units of similar capacity."

An additional possible disadvantage is that the cost of installing mini splits can be higher than some systems. However, lower operating costs and rebates or other financial incentives—offered in some areas—can help offset the initial expense.

Central (Ducted) Air Conditioning

Central (ducted) air conditioning offers whole-house or large-commercial-space cooling, and often offers moderate multi-zone temperature control capability by the addition of air-louver-control boxes.

In central air conditioning, the inside heat-exchanger is typically placed inside the central furnace/AC unit of the forced airheating system which is then used in the summer to distribute chilled air throughout a residence or commercial building.

The heat-exchanger cools the air that is being forced through it by the furnace blower. As the warm air comes in contact with this cool surface the water in the air condenses. By pulling the water molecules from the air. According to the psychometric chart as relative humidity decreases in order to feel cool you will have to lower the temperature even more. A common way to counteract this effect is by installing a whole-home humidifier. Similarly, installing a high efficient system this need to turn the temperature down wont have such and influence on your energy costs.

Multi-split System

Multi-split system is a conventional split system, which is divided into two parts (evaporator and condenser) and allows cooling or heating of several rooms with one external unit. In the outdoor unit of this air conditioner there is a more powerful compressor, ports for connecting several traces and automation with locking valves for regulating the volume of refrigerant supplied to the indoor units located in the room.

Difference between Split System and Multi-split System

Other common types of air conditioning system are multi-split systems, the difference between separate split system and multi-split system in several indoor units. All of them are connected to the main external unit, but the principle of their operation is similar to a simple split-system.

Its unique feature is the presence of one main external unit that connected to several indoor units. Such systems might be the right solution for maintaining the microclimate in several offices, shops, large living spaces. Just few of outdoor units do not worsen the aesthetic appearance of the building.The main external unit can be connected to several different indoor types: floor, ceiling, cassette, etc.

Multi-split System Installation

Before selecting the installation location of air conditioner, several main factors need to be considered. First of all, the direction of air flow from the indoor units should not fall on the place of rest or work area. Secondly, there should not be any obstacles on the way of the airflow that might prevent it from covering the space of the premises as much as possible. The outdoor unit must also be located in an open space, otherwise the heat from the house will not be effectively discharged outside and the productivity of the entire system will drop sharply. It is highly advisable to install the air conditioner units in easily accessible places, for further maintenance during operation.

The main problem when installing a multi-split system is the laying of long refrigerant lines for connecting the external unit to the internal ones. While installing a separate split system, workers try to locate both units opposite to each other, where the length of the line is minimal. Installing a multi-split system creates more difficulties, since some of indoor units can be located far from the outside. The first models of multi-split systems had one common control system that did not allow you to set the air conditioning individually for each room. However, now the market has a wide selection of multi-split systems, in which the functional characteristics of indoor units operate separately from each other.

The selection of indoor units has one restriction - their total power should not exceed the capacity of the outdoor unit. In practice, however, it's very common to see a multi-split system with a total capacity of indoor units greater than the outdoor capacity by at least 20%. But, it is wrong to expect better performance when all indoor units are turned on at the same time, since the total capacity of the whole system is limited by the capacity of the outdoor unit. Simply put, the outdoor unit will distribute all its power to all operating indoor units in such a way that some of the rooms may not have a very comfortable temperature level. However, the calculation of the total power is not simple, since it takes into account not only the nominal power of the units, but also the cooling capacity, heating, dehumidification, humidification, venting, etc.

Portable Units

A portable air conditioner can be easily transported inside a home or office. They are currently available with capacities of about 5,000–60,000 BTU/h (1,500–18,000 W) and with or without electric-resistance heaters. Portable air conditioners are either evaporative or refrigerative.

The compressor-based refrigerant systems are air-cooled, meaning they use air to exchange heat, in the same way as a car radiator or typical household air conditioner does. Such a system dehumidifies the air as it cools it. It collects water condensed from the cooled air and produces hot air which must be vented outside the cooled area; doing so transfers heat from the air in the cooled area to the outside air.

Portable Split System

A portable system has an indoor unit on wheels connected to an outdoor unit via flexible pipes, similar to a permanently fixed installed unit.

Portable Hose System

Hose systems, which can be *monoblock* or *air-to-air*, are vented to the outside via air ducts. The *monoblock* type collects the water in a bucket or tray and stops when full. The *air-to-air* type re-evaporates the water and discharges it through the ducted hose and can run continuously.

A single-hose unit uses air from within the room to cool its condenser, and then vents it outside. This air is replaced by hot air from outside or other rooms (due to the negative pressure inside the room), thus reducing the unit's overall efficiency.

Modern units might have a coefficient of performance of approximately 3 (i.e., 1 kW of electricity will produce 3 kW of cooling). A dual-hose unit draws air to cool its condenser from outside instead of from inside the room, and thus is more effective than most single-hose units. These units create no negative pressure in the room.

Portable Evaporative System

Evaporative coolers, sometimes called "swamp coolers", do not have a compressor or condenser. Liquid water is evaporated on the cooling fins, releasing the vapor into the cooled area. Evaporating water absorbs a significant amount of heat, the latent heat of vaporisation, cooling the air. Humans and animals use the same mechanism to cool themselves by sweating.

Evaporative coolers have the advantage of needing no hoses to vent heat outside the cooled area, making them truly portable. They are also very cheap to install and use less energy than refrigerative air conditioners.

Uses

Air-conditioning engineers broadly divide air conditioning applications into *comfort* and *process* applications.

Comfort Applications

An array of air conditioners outside a commercial office building

Comfort applications aim to provide a building indoor environment that remains relatively constant despite changes in external weather conditions or in internal heat loads.

Air conditioning makes deep plan buildings feasible, for otherwise they would have to be built narrower or with wells so that inner spaces received sufficient outdoor air via natural ventilation. Air conditioning also allows buildings to be taller, since wind speed increases significantly with altitude making natural ventilation impractical for very tall buildings. Comfort applications are quite different for various building types and may be categorized as:

- Commercial buildings, which are built for commerce, including offices, malls, shopping centers, restaurants, etc.,

- High-rise residential buildings, such as tall dormitories and apartment blocks,

- Industrial spaces where thermal comfort of workers is desired,

- Cars, aircraft, boats, which transport passenger or fresh goods,

- Institutional buildings, which includes government buildings, hospitals, schools, etc.,

- Low-rise residential buildings, including single-family houses, duplexes, and small apartment buildings,

- Sports stadiums, such as the University of Phoenix Stadium and in Qatar for the 2022 FIFA World Cup.

Women have, on average, a significantly lower resting metabolic rate than men. Using inaccurate metabolic rate guidelines for air conditioning sizing can result in oversized and less efficient equipment, and setting system operating setpoints too cold can result in reduced worker productivity.

In addition to buildings, air conditioning can be used for many types of transportation, including automobiles, buses and other land vehicles, trains, ships, aircraft, and spacecraft.

Domestic Usage

Typical residential central air conditioners in North America

Air conditioning is common in the US, with 88% of new single-family homes constructed in 2011 including air conditioning, ranging from 99% in the South to 62% in the West. In Canada, air conditioning use varies by province. In 2013, 55% of Canadian households reported having an air conditioner, with high use in Manitoba (80%), Ontario (78%), Saskatchewan (67%), and Quebec (54%) and lower use in Prince Edward Island (23%), British Columbia (21%), and Newfoundland and Labrador (9%). In Europe, home air conditioning is generally less common. Southern Europeancountries such as Greece have seen a wide proliferation of home air-conditioning units in recent years. In another southern European country, Malta, it is estimated that around 55% of households have an air conditioner installed. In India AC sales have dropped by 40% due to higher costs and stricter energy efficiency regulations.

Process Applications

Process applications aim to provide a suitable environment for a process being carried out, regardless of internal heat and humidity loads and external weather conditions. It is the needs of the process that determine conditions, not human preference. Process applications include these:

- Chemical and biological laboratories.

- Cleanrooms for the production of integrated circuits, pharmaceuticals, and the like, in which very high levels of air cleanliness and control of temperature and humidity are required for the success of the process.

- Environmental control of data centers.

- Facilities for breeding laboratory animals. Since many animals normally reproduce only in spring, holding them in rooms in which conditions mirror those of spring all year can cause them to reproduce year-round.

- Food cooking and processing areas.

- Hospital operating theatres, in which air is filtered to high levels to reduce infection risk and the humidity controlled to limit patient dehydration. Although temperatures are often in the comfort range, some specialist procedures, such as open heart surgery, require low temperatures (about 18° C, 64° F) and others, such as neonatal, relatively high temperatures (about 28° C, 82° F).

- Industrial environments.

- Mining.

- Nuclear power facilities.

- Physical testing facilities.

- Plants and farm growing areas.

- Textile manufacturing.

In both comfort and process applications, the objective may be to not only control temperature, but also humidity, air quality, and air movement from space to space.

Health Effects

In hot weather, air conditioning can prevent heat stroke, dehydration from excessive sweating and other problems related to hyperthermia. Heat waves are the most lethal type of weather phenomenon in developed countries. Air conditioning (including filtration, humidification, cooling and disinfection) can be used to provide a clean, safe, hypoallergenic atmosphere in hospital operating rooms and other environments where proper atmosphere is critical to patient safety and well-being. It is sometimes recommended for home use by people with allergies.

Poorly maintained water cooling towers can promote the growth and spread of microorganisms, such as *Legionella pneumophila*, the infectious agent responsible for Legionnaires' disease, or thermophilic actinomycetes. As long as the cooling tower is kept clean (usually by means of a chlorine treatment), these health hazards can be avoided or reduced. Excessive air conditioning can have a negative effect on skin, causing it to dry out, and can also cause dehydration.

Environmental Impacts

Power Consumption and Efficiency

Innovation in air conditioning technologies continues, with much recent emphasis placed on energy efficiency. Production of the electricity used to operate air conditioners has an environmental impact, including the release of greenhouse gases.

Cylinder unloaders are a method of load control used mainly in commercial air conditioning systems. On a semi-hermetic (or open) compressor, the heads can be fitted with unloaders which remove a portion of the load from the compressor so that it can run better when full cooling is not needed. Unloaders can be electrical or mechanical.

According to a 2015 government survey, 87% of the homes in the United States use air conditioning and 65% of those homes have central air conditioning. Most of the homes with central air conditioning have programmable thermostats, but approximately two-thirds of the homes with central air do not use this feature to make their homes more energy efficient.

Lower-energy Alternatives

Alternatives to continual air conditioning can be used with less energy, lower cost, and with less environmental impact. These include:

- In large commercial buildings, making windows able to be opened by occupants when the air outside is cool enough to be comfortable
- Setting thermostats to around 82° F (28° C) and allowing workers to wear more climate-appropriate clothing, such as polo shirts and Bermuda shorts. This approach has worked for the Cool Biz campaign in Japan.
- Passive cooling techniques, such as:
  - Natural ventilation under and through buildings
  - Operating windows to induce a stack effect breeze
  - Letting in cool air at night and closing windows during the day
  - Operating shades to reduce solar gain
  - Building slightly underground, to take advantage of unpowered conduction and geothermal mass
  - Placement of trees, architectural shades, windows (and using window coatings) to reduce solar gain
  - Thermal insulation placed to prevent heat from entering
  - Light-colored building materials reflect away more incoming infrared radiation.
- Using a fan if the air is below body temperature
- Swamp coolers in hot but dry weather
- Using a geothermal heat pump or ground-coupled heat exchanger
- Using naturally cooler basement rooms more
- Taking a siesta during the hottest part of the day
- Sleeping outside on a porch or roof.

Window Air-conditioning System

Window air conditioner is sometimes referred to as room air conditioner as well. It is the simplest form of an air conditioning system and is mounted on windows or walls. It is a single unit that is assembled in a casing where all the components are located.

This refrigeration unit has a double shaft fan motor with fans mounted on both sides of the motor. One at the evaporator side and the other at the condenser side.

The evaporator side is located facing the room for cooling of the space and the condenser side outdoor for heat rejection. There is an insulated partition separating this two sides within the same casing.

Front Panel

The front panel is the one that is seen by the user from inside the room where it is installed and has a user interfaced control be it electronically or mechanically. Older unit usually are of mechanical control type with rotary knobs to control the temperature and fan speed of the air conditioner.

The newer units come with electronic control system where the functions are controlled using remote control and touch panel with digital display.

The front panel has adjustable horizontal and vertical(some models) louvers where the direction of air flow are adjustable to suit the comfort of the users.

The fresh intake of air called VENT (ventilation) is provided at the panel in the event that user would like to have a certain amount of fresh air from the outside.

The mechanical type is usually lower in price compared to the electronic type. If you just want to cool the room and are not too particular about aesthetic or additional functions, the mechanical type will do the work.

Indoor Side Components

The indoor parts of a window air conditioner include:

- Cooling Coil with a air filter mounted on it. The cooling coil is where the heat exchange happen between the refrigerant in the system and the air in the room.

- Fan Blower is a centrifugal evaporator blower to discharge the cool air to the room.

- Capillary Tube is used as an expansion device. It can be noisy during operation if installed too near the evaporator.

- Operation Panel is used to control the temperature and speed of the blower fan. A thermostat is used to sense the return air temperature and another one to monitor the temperature of the coil. Type of control can be mechanical or electronic type.

- Filter Drier is used to remove the moisture from the refrigerant.

- Drain Pan is used to contain the water that condensate from the cooling coil and is discharged out to the outdoor by gravity.

A typical window unit.

Outdoor Side Components

The outdoor side parts include:

- Compressor is used to compress the refrigerant.

- Condenser Coil is used to reject heat from the refrigeration to the outside air.

- Propeller Fan is used in air-cooled condenser to help move the air molecules over the surface of the condensing coil.

- Fan Motor is located here. It has a double shaft where the indoor blower and outdoor propeller fan are connected together.

Working of Window Air Conditioner

The working of window air conditioner can be explained by separately considering the two cycles of air: room air cycle and the hot air cycle. The compartments of the room and hot air are separated by an insulated partition inside the body of the air conditioner. The setting of thermostat and its working has also been explained in the discussions below.

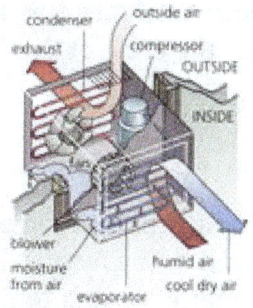

Working of Window AC

- Room Air Cycle

The air moving inside the room and in the front part of the air conditioner where the cooling coil is located is considered to be the room air. When the window AC is started the blower starts immediately and after a few seconds the compressor also starts. The evaporator coil or the cooling gets cooled as soon as the compressor is started.

The blower behind the cooling coil starts sucking the room air, which is at high temperature and also carries the dirt and dust particles. On its path towards the blower, the room air first passes through the filter where the dirt and dust particles from it get removed.

The air then passes over the cooling coil where two processes occur. Firstly, since the temperature of the cooling coil is much lesser than the room air, the refrigerant inside the cooling coil absorbs the heat from the air. Due to this the temperature of the room air becomes very low, that is the air becomes chilled.

Secondly, due to reduction in the temperature of the air, some dew is formed on the surface of the cooling coil. This is because the temperature of the cooling coil is lower than the dew point temperature of the air. Thus the moisture from the air is removed so the relative humidity of the air reduces. Thus when the room air passes over the cooling coil its temperature and relative humidity reduces.

This air at low temperature and low humidity is sucked by the blower and it blows it at high pressure. The chilled air then passes through small duct inside the air conditioner and it is then thrown outside the air conditioner through the opening in the front panel or the grill. This chilled air then enters the room and chills the room maintaining low temperature and low humidity inside the room.

The cool air inside the room absorbs the heat and also the moisture and so its temperature and moisture content becomes high. This air is again sucked by the blower and the cycle repeats. Some outside air also gets mixed with this room air. Since this air is sent back to the blower, it is also called as the return room air. In this way the cycle of this return air or the room air keeps on repeating.

Hot Air Cycle

The hot air cycle includes the atmospheric air that is used for cooling the condenser. The condenser of the window air conditioner is exposed to the external atmosphere. The propeller fan located behind the condenser sucks the atmospheric at high temperature and it blows the air over the condenser.

The refrigerant inside the condenser is at very high temperature and it has to be cooled to produce the desired cooling effect. When the atmospheric air passes over the condenser, it absorbs the heat from the refrigerant and its temperature increases. The atmospheric air is already at high temperature and after absorbing the condenser heat, its temperature becomes even higher. The person standing behind the condenser of the window AC can clearly feel the heat of this hot air. Since the temperature of this air is very high, this is called as hot air cycle.

this indoor cabinet also contains a furnace or the indoor part of a heat pump. The air conditioner's evaporator coil is installed in the cabinet or main supply duct of this furnace or heat pump. If your home already has a furnace but no air conditioner, a split-system is the most economical central air conditioner to install.

In a packaged central air conditioner, the evaporator, condenser, and compressor are all located in one cabinet, which usually is placed on a roof or on a concrete slab next to the house's foundation. This type of air conditioner also is used in small commercial buildings. Air supply and return ducts come from indoors through the home's exterior wall or roof to connect with the packaged air conditioner, which is usually located outdoors. Packaged air conditioners often include electric heating coils or a natural gas furnace. This combination of air conditioner and central heater eliminates the need for a separate furnace indoors.

Choosing or Upgrading Your Central Air Conditioner

Central air conditioners are more efficient than room air conditioners. In addition, they are out of the way, quiet, and convenient to operate. To save energy and money, you should try to buy an energy-efficient air conditioner and reduce your central air conditioner's energy use. In an average air-conditioned home, air conditioning consumes more than 2,000 kilowatt-hours of electricity per year, causing power plants to emit about 3,500 pounds of carbon dioxide and 31 pounds of sulfur dioxide.

If you are considering adding central air conditioning to your home, the deciding factor may be the need for ductwork.

If you have an older central air conditioner, you might choose to replace the outdoor compressor with a modern, high-efficiency unit. If you do so, consult a local heating and cooling contractor to assure that the new compressor is properly matched to the indoor unit. However, considering recent changes in refrigerants and air conditioning designs, it might be wiser to replace the entire system.

Today's best air conditioners use 30% to 50% less energy to produce the same amount of cooling as air conditioners made in the mid 1970s. Even if your air conditioner is only 10 years old, you may save 20% to 40% of your cooling energy costs by replacing it with a newer, more efficient model.

Proper sizing and installation are key elements in determining air conditioner efficiency. Too large a unit will not adequately remove humidity. Too small a unit will not be able to attain a comfortable temperature on the hottest days. Improper unit location, lack of insulation, and improper duct installation can greatly diminish efficiency.

When buying an air conditioner, look for a model with a high efficiency. Central air conditioners are rated according to their seasonal energy efficiency ratio (SEER). SEER indicates the relative amount of energy needed to provide a specific cooling output. Many older systems have SEER ratings of 6 or less.

Working of Central Air Conditioner

When the thermostat signals the air-conditioning system to lower air temperature, a whole sequence of events begins.

First, the air-handling unit kicks on, drawing room air in from various parts of the house through return-air ducts. This air is pulled through one or more filters, where airborne particles such as dust and lint are removed—in fact, sophisticated filters may remove microscopic pollutants as well. Then the air is routed to air-supply ductwork through which the blower pushes it back to the rooms.

But how does the evaporator coil get cold in the first place? That is where refrigeration principles come into play.

Every air conditioner has three main parts: a condenser, an evaporator, and a compressor. With a typical "split system," the condenser and the compressor are located in an outdoor unit. The evaporator is mounted on or in the air-handling unit, which is often a forced-air furnace. With a "package system," all of the components are combined in a single outdoor unit that may be located on the ground or on the roof.

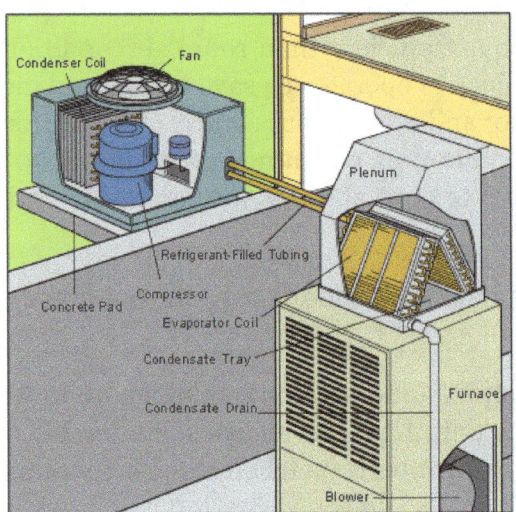

A central air conditioner cools with an outdoor compressor and condenser
coil connected to an indoor furnace fitted with an evaporator coil

Refrigerant circulates through copper tubing that runs between the evaporator and the condenser. This refrigerant receives and releases heat as it raises and lowers in temperature, changing from liquid to gas and then back to liquid. The refrigerant is especially cold when it begins to circulate through the indoor coil.

As the air handler pushes warm air across the coil, the refrigerant absorbs so much heat from the air that it turns into vapor. As a vapor, it travels to the outdoor compressor, which pressurizes it and moves it through the outdoor coil. There it jettisons the heat through coils or thin metal "fins." A fan in the compressor also helps to dissipate the heat. The refrigerant then passes through an expansion device that converts it to a low-pressure, low-temperature liquid, which returns to the indoor coil. And so the cycle goes.

Package Air-conditioning System

Package air conditioner is a bigger version of the window air conditioner. However, unlike window air conditioner or PTAC units, it has a higher cooling or heating capacity and is usually able to cool an entire house or a commercial building. The nominal capacities ranges from 3 tonne to 15 tonne.

The conditioned air are transferred to the space to be conditioned through ducting which is usually hidden in the ceiling and wall of the building. The unit is placed outside the house, a special room in a building or even on top of a roof. This unit is factory assembled and skilled technicians are needed to install this type of unit.

Protection devices such as High/Low pressure switch, overload relays for all motors, water flow and air flow switches are included in the units. The compressor have winding protection thermostat built into the winding to disconnect the circuit in the event of overheating.

There is an interlocking circuit with the evaporator fan motor starter to ensure that the compressor can only start if the fan motor is running.

Components

This self contained unit is assembled in a casing where all the air conditioning componenets are housed. They include:

- Compressor usually hermetic or semi-hermetic type for operation on 380/400 Volts 3 Phase is used.

- Water-cooled or Air-cooled condenser.

- Electrical Panel.

- Thermostatic Expansion Valve.

- Air Filter.

- Front Panel & Return Air Grill.

- Evaporator Coil.

- Evaporator Fan And Housing.

- Heating and Humidifying Components may be included in the unit. Dehumidification is needed for cooling mode during summer and Humidification for heating mode during winter.

Rooftop Air-Cooled Package Air Conditioner

Package Air Conditioner Condenser Type

The condenser used in a package air conditioner can be air-cooled or water-cooled type. Air-cooled type is usually for capacity below 5 tonne and water-cooled for capacity above 5 tonne. The capacity of air-cooled type is lower than that of water-cooled condenser using the same compressor.

The water-cooled type can be completely factory assembled, tested and charged with refrigerant before being installed in the field. This is advantages because less man power is needed in the field to do the installation hence cost saving.

The air-cooled type cannot be factory assembled or charged as the laying of refrigerant piping, pressure testing, charging and evacuation have to be carried out in the field.

Air Quantities

In the air distribution, centrifugal fans are usually provided at the evaporator side which can develop higher static pressure. The higher the static pressure, the better the air throw to the room is. Air distribution is done through ducts and grills. The air quantitites are usually in the range of 10-11.3 cubic meter/min (350-400 cfm) per tonne.

In critical application such as computer room which need higher rate of air flow, higher capacity fans with air flow up to 550 cfm are provided. This usually need to be factored in during the design of the system.

Two Refrigeration Systems

In some units, two independent refrigeration systems are included in the package. In this case, the cooling coils have a common tube sheets though the two systems are independent. The tubes of the coil are entwined.

Two separate water-cooled condensers or a condenser with an internal partition that forms two independent condensers are used. The water side is common to both the independent condenser portions.

It is a common practice to have one compressor running all the time with the other compressor control by a single-stage thermostat. In this case, the capacity can be controlled either to 50% or 100%. However, newer units have two-stage thermostat that can control the capacity from 0%, 50% or 100%.

Working of Packaged System

Operation depends on configuration, but packaged systems typically heat and cool your home the same way their stand-alone counterparts do. The ducting with a single cabinet system is slightly different. The duct work is attached to the system rather than connecting to various components in your home.

Packaged System Air Condition Component

- By using electricity as its power source, the unit's internal components cycle the refrigerant.

- Warm air is pulled in by a fan and then passes over the cold evaporator coil, cooling it in the process.

- The cooled, dehumidified air is pushed through ducts to the various spaces inside your home.

Package System Heating Component

- Packaged Air Conditioners: In addition to the typical cooling feature associated with an air conditioner, packaged air conditioners are capable of producing limited heat with heat strip elements. With electricity as the fuel source, the heat strips are warmed, and the air is heated as it flows over the strips.The warm air then travels through ducting to increase the interior temperature of your home. This type of heating component is mainly used in warmer climates where heat is only used occasionally.

- Packaged Heat Pumps: The heat pump transfers heat by reversing the refrigeration cycle used by a typical air conditioner. Through a cycle of evaporation and condensation, the indoor coils are heated, and the air is pushed over the warm coils. From there, the warmed air is blown through the ductwork to increase the temperature in the interior rooms of your home.

- Packaged Gas-electric: The heating component of a packaged gas-electric system is a gas furnace. The heating portion of the system uses natural gas or propane to combust inside the heat exchanger, creating heat. As cool air from the interior spaces is pulled in through the return ducting, the blower motor then blows the air over and through the hot heat exchanger, heating the air. The warm air is then circulated throughout the home through the ductwork.

- Packaged Dual-Fuel: Your dual-fuel packaged system has two heating options, a heat pump or a gas furnace. When installed and configured correctly, your dual fuel system can determine whether it's more economical to heat your home using electricity or gas. When moderate heating is required, the heat pump automatically reverses from the air condition mode to provide warm air. When temperatures fall further, the system uses the gas furnace to provide reliable, consistent heat.

Benefits

- Space efficiency – Unlike split-system units, all components of a complete heating and cooling system are contained in one location, making packaged units ideal for situations in which indoor space is at a premium.

- Energy-efficient heating and cooling performance – All Goodman brand packaged units offer 13 SEER or higher cooling performance.

District Cooling System

District cooling means the centralized production and distribution of cooling energy. Chilled water is delivered via an underground insulated pipeline to office, industrial and residential buildings to cool the indoor air of the buildings within a district. Specially designed units in each building then use this water to lower the temperature of air passing through the building's air conditioning system.

The output of one cooling plant is enough to meet the cooling-energy demand of dozens of buildings. District cooling can be run on electricity or natural gas, and can use either regular water or seawater. Along with electricity and water, district cooling constitute a new form of energy service.

District cooling is measured in refrigeration ton which is equivalent to 12000 BTU's per hour. Refrigeration Ton is the unit measure for the amount of heat removed. Refrigeration Ton is defined as the heat absorbed by one ton of ice (2000 pounds) causing it to melt completely by the end of one day (24 hours).

District cooling systems can replace any type of air conditioning system, but primarily compete with air-cooled reciprocating chiller systems serving large buildings which consume large amounts of electricity. This air-conditioning system is subject to a difficult operating environment, including extreme heat, saline humidity and windborne sand. Over time, performance, efficiency and reliability suffer, leading to significant maintenance costs and ultimately to equipment replacement.

Benefits of District Cooling System

There are technical, commercial and environmental benefits by using a DCS compare to a decentralized cooling system:

- Guaranteeing chilled water tariffs and performance to End-users
- Reducing overall electricity infrastructure cost and capital expenditure
- Reducing energy consumption
- Reducing operational and maintenance costs
- Increasing reliability and efficiency by a 24/7 professional O&M team
- Increasing comfort and safety
- Enhancing real-estate net space and urban landscape
- Enhancing a sustainable development
- Allowing building managers to focus on core business.

Chilled Beam System

An alternative to a conventional VAV system, this chilled beam cooling device separates ventilation and dehumidification. It's made of copper tubing and bonded to aluminum fins; it's housed in a sheet metal enclosure and is usually placed at ceiling level.

Chilled beam systems are available in three variations: passive, active and integrated/multi-service beams. The difference between passive and active beams revolves around the way airflow and fresh air are brought into the space. Integrated/multi-service beams are chilled beam systems that include lighting, speakers, sprinkler openings, cable pathways, etc.

Working of Chilled Beam System

With goals of eliminating excessive fan energy and reliance on reheat, a chilled beam acts as a radiator chilled by recirculated water. Warm air rises and is cooled by the chilled beam; once it's cooled, the air falls back to the floor, where the cycle starts over. Passive chilled beams require ventilation air to be delivered by a separate air-handling system; in an active beam, the ventilation air is delivered to the beam by a central air-handling system via ductwork.

Advantages and Disadvantages

The primary advantage of the chilled beam system is its lower operating cost. For example, because the temperature of cooled water is higher than the temperature of cooled air, but it delivers the same cooling ability, the costs of the cooled water system are lower. Because cooling and heating of air are no longer linked to the delivery of air, buildings also save money by being able to run fewer air circulation fans and at lower speeds. One estimate places the amount of air handled at 25 to 50 percent less using chilled beam systems. By being able to target the delivery of clean outdoor air where it is needed (rather than injecting it into the entire system and heating or cooling it), there is a reduced need to treat large amounts of outdoor air (also saving money). In one case, the Genomic Science Building at the University of North Carolina at Chapel Hill lowered its HVAC costs by 20 percent with an active chilled beam system. This is a typical energy cost savings. Chilled beam systems also have some advantages in that they are almost noiseless, require little maintenance, and are highly efficient. Traditional fan-driven HVAC systems create somewhat higher air velocities, which some people find uncomfortable. Chilled beam HVAC systems also require less ceiling space than forced-air HVAC systems, which can lead to lower building heights and higher ceilings. Since they do not require high forced air flows, chilled beam systems also require reduced air distribution duct networks (which also helps to lower cost).

Chilled beam systems are not a panacea. Additional ductwork may be needed to meet minimum outdoor air requirements. Both types of chilled beam systems are less effective at heating than cooling, and supplementary heating systems are often needed. Chilled beam systems cannot be used alone in buildings where the ceilings are higher than 2.7 metres (8.9 ft), because the air will not properly circulate. A forced-air circulation system must be employed in such cases. If the water temperature is too low or humidity is high, condensation on the beam can occur—leading to a problem known as "internal rain." (In some cases, drier outside air can be mixed with the wetter inside air to reduce interior humidity levels while maintaining system performance.) Chilled beam systems are not recommended for areas with high humidity (such as theaters, gymnasiums, or cafeterias). Because they are less effective at cooling, passive chilled beam systems are generally ill-suited for semi-tropical and tropical climates. Hospitals generally cannot use chilled beam systems because of restrictions on using recirculated air. Chilled beam systems are also known to cause noticeable air circulation which can make some people uncomfortable. (Passive air deflection devices can help disrupt these air patterns, alleviating the problem.) Some designers have found that enlarging the ducts around active chilled beam systems to increase air circulation causes echoes in working areas and amplifies the sound of water moving through the pipes to noticeable levels.

Installation and Adoption

Active chilled beams are mounted in a suspended ceiling and then anchored to the overhead structure, because T-bar ceilings cannot support the typical operating weight of a chilled beam. They are generally 1 to 2 feet (0.30 to 0.61 m) wide, and require less than 1 feet (0.30 m) of overhead space. A typical 2-feet (0.61 m) wide chilled beam system generally weighs about 15 pounds (6.8 kg) per 1 feet (0.30 m) length of the beam. Chilled beams are generally installed so that the center of each beam is no more than 3 metres (9.8 ft) from the center of the next beam. Some architects and end-users dislike the beams because they do not cover the entire ceiling so ducts, wiring, and other infrastructure can be seen. Some designers have installed one chilled beam system around the building perimeter (where temperature differences can be the greatest) and another in the interior of the building, to better control temperature throughout the structure. Higher system performance may be obtained by increasing the static pressure of the air in the building. The systems generally need little cleaning (vacuuming of dirt and dust from the fins every five years).

Heat Pump and Refrigeration Cycle

Heat pumps are devices that operate in a cycle similar to the vapor-compression refrigerator cycle.

In its most basic form, a vapor-compression refrigeration system consists of an evaporator, a compressor, a condenser, a throttling device which is usually an expansion valve or capillary tube and the connecting tubing. The working fluid is the refrigerant, such as freon or ammonia, which goes through a thermodynamic cycle.

The thermodynamic cycle is shown schematically in the figure below. Four important processes take place during the cycle. First, heat (Q_1) is transferred to the refrigerant in the evaporator from stations 4 to 1 in figures below, where both its pressure and temperature are lower than a thermal source such as air, water, or the ground. Evaporation of the refrigerant occurs from a liquid to a saturated vapor, theoretically at a constant pressure. In practice, however, a pressure drop is associated with fluid flow and heat transfer through the evaporator. Secondly, work (W) is done on the refrigerant as saturated vapor at low pressure and temperature enters the compressor and undergoes adiabatic compression (from 1 to 2). The result is a compressed refrigerant vapor at high pressure and temperature at the compressor outlet. Third, heat (Q_y) is transferred from the hot vapor in the condenser (from 2 to 3), where its pressure and temperature are higher than a thermal sink that is at a higher temperature than the source. Condensation from a vapor to a saturated liquid occurs in the condenser, theoretically at constant pressure, but again in practice a pressure drop occurs for the same reasons as in the evaporator. The refrigerant leaves the condenser as saturated liquid. The last process before the refrigerant reenters the evaporator is the throttling of the refrigerant through the expansion valve or capillary tube from 3 to 4. During this process the pressure drop is adiabatic, resulting in a decreased refrigerant pressure and temperature. Usually the refrigerant enters as a liquid and leaves as a mixture of liquid and vapor.

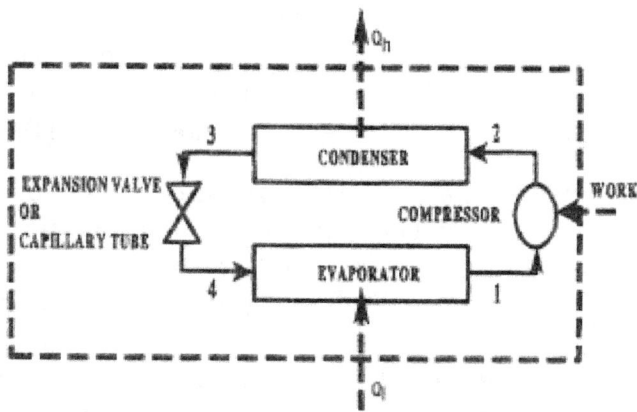

A simple vapor-compression refrigeration cycle

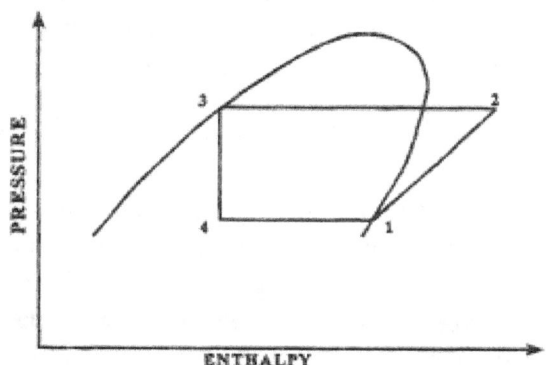

Thermodynamic cycle of the vapor-refrigeration cycle

The basic heat pump cycle is identical to the vapor-compression refrigeration cycle the only difference between a heat pump and a refrigerator being their basic functions. A refrigeration system cools the external fluid flowing through the evaporator, whereas a heat pump heats the external fluid flowing through the condenser. The main difference between a refrigerator and a heat pump is in the manner of operation regarding cooling or heating. If the application is cooling then you would be interested in the cooling aspect, Q_l in figure above occuring over the evaporator, and the cooling device will, for example, be called refrigerator, air conditioner, chiller, crycooler, etc. On the other hand, if the application is heating, then you would be interested in the heating aspect, Q_y in figure above, occuring over the condenser, and the heating device will be called a heat pump.

Heat pumps are mostly used for heating water and air. Water can be heated for swimming pools and household purposes by using ambient air (a so called air-to-water heat pump) and air is usually heated during winter for space heating inside houses, buildings, factories, etc. also by using ambient air as the source (air-to-air heat pump). The ground can, however, also be used as a source for space heating. The heat pump will then be called a ground-coupled, ground source or ground-to-air heat pump. Another source of heat is water. Here, the evaporator is placed in a borehole, pond or lake for space heating. This type of heat pump is called a water-to-air heat pump. In the industry water-to-water heat pumps are used, for example, to produce hot and cold water simultaneously. These types of systems are also called dual-purpose systems.

Dual-purpose systems have been designed to be applied in both heating and cooling capacities at the same time. In a reverse cycle system, the functions of evaporator and condenser can be reversed. Both "hot" and "cold" can thus be delivered to the same thermal reservoir at different times. Dual systems are also called heat pumps, although the focus is not solely on their heating capacities.

Although the devices have different names, two features are common in both. First, the same basic refrigeration cycle takes place and second both are heat-pumping systems. Heat is "pumped" from a thermal source at low temperature to a thermal sink at a higher temperature. These heat-pumping systems or heat pumps have two major advantages over conventional technology. First, depending on the application, more than one unit of heating per unit of energy input required can often be delivered, i.e., the coefficient of performance (COP) value is greater than one. Usually, for every one kilowatt of power required by the compressor more than one kilowatt of heating capacity is available at the condenser. In most practical applications the coefficient of performance is between two and six for a heat pump. To heat a swimming pool, the COP may be as high as six; whereas in a hot water system, where water is heated to a temperature of 55° C, the COP may be as low as three. The coefficient of performance (COP) of a heat pump is defined as,

$$COP = Q_h \, (\text{energy sought}) \, / \, W \, (\text{energy that costs}) = Q_h / (Q_h - Q_1).$$

A COP value of four would mean an energy saving of 75%. Heat pumps therefore offer the advantage of energy conservation and lowered costs compared to other methods of heating. Another advantage of heat pumps is that heat-pumping devices may be either heat-actuated or work-actuated. Heat-actuated heat pumps allow for use of lower-grade thermal energy, which in other cases might often remain unused. Heat pumps therefore may also help mitigate thermal pollution and environmental concerns.

The advantages of heat pumps have long been recognized. Research into improving performance, reliability, energy-efficiency, and environmental impact has been an ongoing concern for industrial, governmental, and academic organizations. Studies have centered on advanced cycle design for both heat- and work-actuated systems, improved components (including choice of refrigerant), and use in a wider range of applications. Specific areas of activity have included: a search for refrigerant replacements, advanced mobile air conditioning for transportation-applications, advanced vapor-compression technology, absorption heat pumps, ground-coupled (or geothermal) heat pumps and air cycle heat pumps.

The Carnot cycles that have been drawn are based on ideal gas behavior. For different working media, however, they will look different. We will see an example when we discuss two-phase situations. What is the same whatever the medium is the efficiency for all Carnot cycles operating between the same two temperatures.

Refrigerator Hardware

Typically the thermodynamic system in a refrigerator analysis will be a working fluid, a refrigerant, that circulates around a loop, as shown in figure. The internal energy (and temperature) of the refrigerant is alternately raised and lowered by the devices in the loop. The working fluid is colder than the refrigerator air at one point and hotter than the surroundings at another point. Thus heat will flow in the appropriate direction, as shown by the two arrows in the heat exchangers.

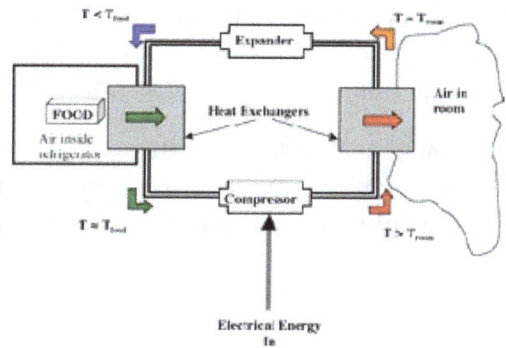

a domestic refrigerator

Starting in the upper right hand corner of the diagram, we describe the process in more detail. First the refrigerant passes through a small turbine or through an expansion valve. In these devices, work is done by the refrigerant so its internal energy is lowered to a point where the temperature of the refrigerant is lower than that of the air in the refrigerator. A heat exchanger is used to transfer energy from the inside of the refrigerator to the cold refrigerant. This lowers the internal energy of the inside and raises the internal energy of the refrigerant. Then a pump or compressor is used to do work on the refrigerant, adding additional energy to it and thus further raising its internal energy. Electrical energy is used to drive the pump or compressor. The internal energy of the refrigerant is raised to a point where its temperature is hotter than the temperature of the surroundings. The refrigerant is then passed through a heat exchanger (often coils at the back of the refrigerator) so that energy is transferred from the refrigerant to the surroundings. As a result, the internal energy of the refrigerant is reduced and the internal energy of the surroundings is increased. It is at this point where the internal energy of the contents of the refrigerator and the energy used to drive the compressor or pump are transferred to the surroundings. The refrigerant then continues on to the turbine or expansion valve, repeating the cycle.

Evaporative Cooler

In low-humidity areas, evaporating water into the air provides a natural and energy-efficient means of cooling. Evaporative coolers, also called swamp coolers, rely on this principle, cooling outdoor air by passing it over water-saturated pads, causing the water to evaporate into it. The 15°- to 40° F-cooler air is then directed into the home, and pushes warmer air out through windows.

Evaporative Cooler

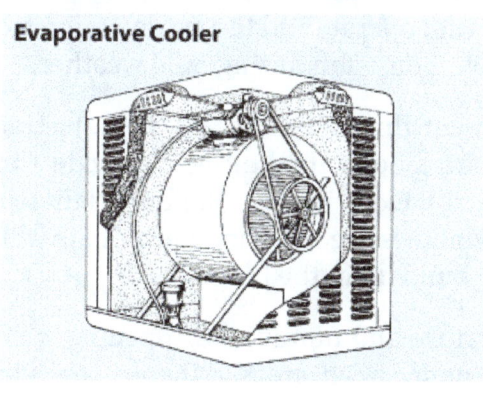

When operating an evaporative cooler, windows are opened part way to allow warm indoor air to escape as it is replaced by cooled air. Unlike central air conditioning systems that recirculate the same air, evaporative coolers provide a steady stream of fresh air into the house.

Evaporative coolers cost about one-half as much to install as central air conditioners and use about one-quarter as much energy. However, they require more frequent maintenance than refrigerated air conditioners and they're suitable only for areas with low humidity.

Sizing and Selection

Evaporative coolers are rated by the cubic feet per minute (cfm) of air that they deliver to the house. Most models range from 3,000 to 25,000 cfm. Manufacturers recommend providing enough air-moving capacity for 20 to 40 air changes per hour, depending on climate.

Installation

Evaporative coolers are installed in one of two ways: the cooler blows air into a central location, or the cooler connects to ductwork, which distributes the air to different rooms. Central-location installations work well for compact houses that are open from room to room. Ducted systems are required for larger houses with hallways and multiple bedrooms.

Most people install down-flow evaporative coolers on the roofs of their houses. However, many experts prefer to install ground-mounted horizontal units, which feature easier maintenance and less risk of roof leaks.

Small horizontal-flow coolers are installed in windows to cool a room or section of a home. These portable evaporative coolers work well in moderate climates, but may not be able to cool a room adequately in hot climates. Room evaporative coolers are becoming more popular in areas of the western United States with milder summer weather. They can reduce the temperature in a single room by 5° to 15° F.

Small, portable evaporative coolers on wheels are now available as well. Although the units have the advantage of portability, their cooling ability is limited by the humidity within your home. Generally, these units will provide only a slight cooling effect.

Operation

An evaporative cooler should have at least two speeds and a vent-only option. During vent-only operation, the water pump does not operate and the outdoor air is not humidified. This lets you use the evaporative cooler as a whole-house fan during mild weather.

Control the cooler's air movement through the house by adjusting window openings. Open the windows or vents on the leeward side of the house to provide 1 to 2 square feet of opening for each 1,000 cfm of cooling capacity. Experiment to find the right windows to open and the correct amount to open them. If the windows are open too far, hot air will enter. If the windows are not open far enough, humidity will build up in the home.

One can regulate both temperature and humidity by opening windows in the areas you want to cool, and closing windows in unoccupied areas. Where open windows create a security issue, install up-ducts in the ceiling. Up-ducts open to exhaust warm air into the attic as cooler air comes in from the evaporative cooler. Evaporative coolers installed with up-ducts will need additional attic ventilation.

Optional filters remove most of the dust from incoming air an attractive option for homeowners concerned about allergies. Filters can also reduce the tendency of some coolers to pull water droplets from the pads into the blades of the fan. Most evaporative coolers do not have air filters as original equipment, but they can be fitted to the cooler during or after installation.

Evaporative Cooler Maintenance

Save yourself a lot of work and money by draining and cleaning your evaporative cooler regularly. Build-up of sediment and minerals should be regularly removed. Coolers need a major cleaning every season, and may need routine maintenance several times during the cooling season.

The more a cooler runs, the more maintenance it will need. In hot climates where the cooler operates much of the time, look at the pads, filters, reservoir, and pump at least once a month. Replace the pads at least twice during the cooling season, or as often as once a month during continuous operation.

Some paper and synthetic cooler pads can be cleaned with soap and water or a weak acid according to manufacturer's instructions. Filters should be cleaned when the pads are changed or cleaned. Be sure to disconnect the electricity to the unit before servicing it.

Two-stage Evaporative Coolers

Two-stage evaporative coolers are newer and even more efficient. They use a pre-cooler, more effective pads, and more efficient motors, and don't add as much humidity to the home as single-stage evaporative coolers. Because of their added expense, they are most often used in areas where daytime temperatures frequently exceed 100° F.

Applications

Before the advent of refrigeration, evaporative cooling was used for millennia. A porous earthenware vessel would cool water by evaporation through its walls; frescoes from about 2500 BC show slaves fanning jars of water to cool rooms. A vessel could also be placed in a bowl of water, covered with a wet cloth dipping into the water, to keep milk or butter as fresh as possible.

California ranch house with evaporative cooler box on roof ridgeline

Evaporative cooling is a common form of cooling buildings for thermal comfort since it is relatively cheap and requires less energy than other forms of cooling.

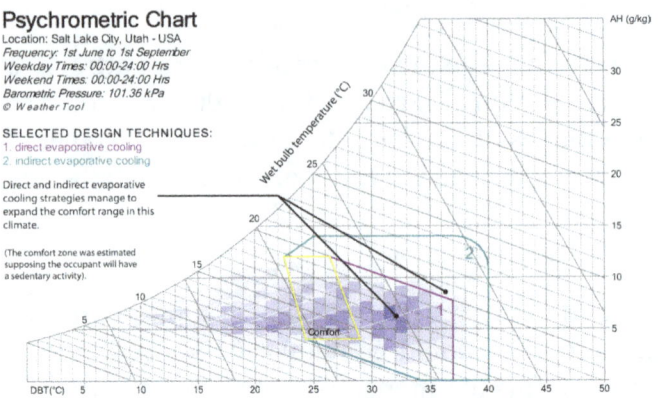

Psychrometric chart example of Salt Lake City

The figure showing the Salt Lake City weather data represents the typical summer climate (June to September). The colored lines illustrate the potential of direct and indirect evaporative cooling strategies to expand the comfort range in summer time. It is mainly explained by the combination of a higher air speed on one hand and elevated indoor humidity when the region permits the direct evaporative cooling strategy on the other hand. Evaporative cooling strategies that involve the humidification of the air should be implemented in dry condition where the increase in moisture content stays below recommendations for occupant's comfort and indoor air quality. Passive cooling towers lack the control that traditional HVAC systems offer to occupants. However, the additional air movement provided into the space can improve occupant comfort.

Evaporative cooling is most effective when the relative humidity is on the low side, limiting its popularity to dry climates. Evaporative cooling raises the internal humidity level significantly, which desert inhabitants may appreciate as the moist air re-hydrates dry skin and sinuses. Therefore, assessing typical climate data is an essential procedure to determine the potential of evaporative cooling strategies for a building. The three most important climate considerations are dry-bulb temperature, wet-bulb temperature, and wet-bulb depression during the summer design day. It is important to determine if the wet-bulb depression can provide sufficient cooling during the summer design day. By subtracting the wet-bulb depression from the outside dry-bulb temperature, one can estimate the approximate air temperature leaving the evaporative cooler. It is important to consider that the ability for the exterior dry-bulb temperature to reach the wet-bulb temperature depends on the saturation efficiency. A general recommendation for applying direct evaporative cooling is to implement it in places where the wet-bulb temperature of the outdoor air does not exceed 22° C (71.6° F). However, in the example of Salt Lake City, the upper limit for the direct evaporative cooling on psychrometric chart is 20° C (68° F). Despite this lower value, this climate is still suitable for this technique.

Evaporative cooling is especially well suited for climates where the air is hot and humidity is low. In the United States, the western/mountain states are good locations, with evaporative coolers prevalent in cities like Denver, Salt Lake City, Albuquerque, El Paso, Tucson, and Fresno. Evaporative air conditioning is also popular and well-suited to the southern (temperate) part of Australia. In dry, arid climates, the installation and operating cost of an evaporative cooler can be much lower than that of refrigerative air conditioning, often by 80% or so. However, evaporative cooling and vapor-compression air conditioning are sometimes used in combination to yield optimal cooling results. Some evaporative coolers may also serve as humidifiers in the heating season. Even in regions that are mostly arid, short periods of high humidity may prevent evaporative cooling from being an effective cooling strategy. An example of this event is the monsoon season in New Mexico and southern Arizona in July and August.

In locations with moderate humidity there are many cost-effective uses for evaporative cooling, in addition to their widespread use in dry climates. For example, industrial plants, commercial kitchens, laundries, dry cleaners, greenhouses, spot cooling (loading docks, warehouses, factories, construction sites, athletic events, workshops, garages, and kennels) and confinement farming (poultry ranches, hog, and dairy) often employ evaporative cooling. In highly humid climates, evaporative cooling may have little thermal comfort benefit beyond the increased ventilation and air movement it provides.

Examples

Trees transpire large amounts of water through pores in their leaves called stomata, and through this process of evaporative cooling, forests interact with climate at local and global scales.

Evaporative cooling is commonly used in cryogenic applications. The vapor above a reservoir of cryogenic liquid is pumped away, and the liquid continuously evaporates as long as the liquid's vapor pressure is significant. Evaporative cooling of ordinary helium forms a 1-K pot, which can cool to at least 1.2 K. Evaporative cooling of helium-3 can provide temperatures below 300 mK. These techniques can be used to make cryocoolers, or as components of lower-temperature cryostats such as dilution refrigerators. As the temperature decreases, the vapor pressure of the liquid also falls, and cooling becomes less effective. This sets a lower limit to the temperature attainable with a given liquid.

Evaporative cooling is also the last cooling step in order to reach the ultra-low temperatures required for Bose–Einstein condensation (BEC). Here, so-called forced evaporative cooling is used to selectively remove high-energetic ("hot") atoms from an atom cloud until the remaining cloud is cooled below the BEC transition temperature. For a cloud of 1 million alkali atoms, this temperature is about 1μK.

Although robotic spacecraft use thermal radiation almost exclusively, many manned spacecraft have short missions that permit open-cycle evaporative cooling. Examples include the Space Shuttle, the Apollo Command/Service Module (CSM), Lunar Module and Portable Life Support System. The Apollo CSM and the Space Shuttle also had radiators, and the Shuttle could evaporate ammonia as well as water. The Apollo spacecraft used sublimators, compact and largely passive devices that dump waste heat in water vapor (steam) that is vented to space. When liquid water is exposed to vacuum it boils vigorously, carrying away enough heat to freeze the remainder to ice that covers the sublimator and automatically regulates the feedwater flow depending on the heat load. The water expended is often available in surplus from the fuel cells used by many manned spacecraft to produce electricity.

Designs

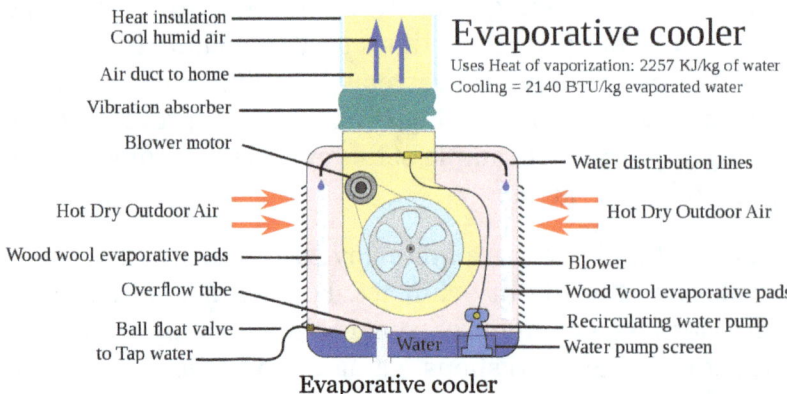

Evaporative cooler

Evaporative cooler

Most designs take advantage of the fact that water has one of the highest known enthalpy of vaporization (latent heat of vaporization) values of any common substance. Because of this, evaporative coolers use only a fraction of the energy of vapor-compression or absorption air conditioning systems. Unfortunately, except in very dry climates, the single-stage (direct) cooler can increase relative humidity (RH) to a level that makes occupants uncomfortable. Indirect and two-stage evaporative coolers keep the RH lower.

Direct Evaporative Cooling

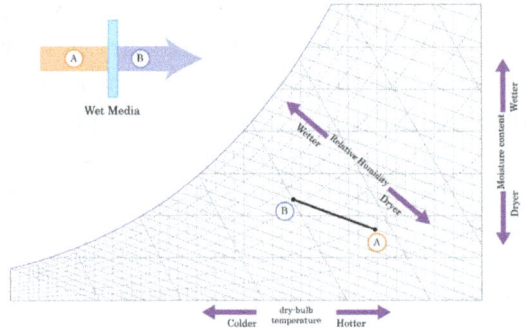

Direct evaporative cooling

Direct evaporative cooling (open circuit) is used to lower the temperature and increase the humidity of air by using latent heat of evaporation, changing liquid water to water vapor. In this process, the energy in the air does not change. Warm dry air is changed to cool moist air. The heat of the outside air is used to evaporate water. The RH increases to 70 to 90% which reduces the cooling effect of human perspiration. The moist air has to be continually released to outside or else the air becomes saturated and evaporation stops.

A *mechanical* direct evaporative cooler unit uses a fan to draw air through a wetted membrane, or pad, which provides a large surface area for the evaporation of water into the air. Water is sprayed at the top of the pad so it can drip down into the membrane and continually keep the membrane saturated. Any excess water that drips out from the bottom of the membrane is collected in a pan and recirculated to the top. Single-stage direct evaporative coolers are typically small in size as they only consist of the membrane, water pump, and centrifugal fan. The mineral content of the municipal water supply will cause scaling on the membrane, which will lead to clogging over the life of the membrane. Depending on this mineral content and the evaporation rate, regular cleaning and maintenance is required to ensure optimal performance. Generally, supply air from the single-stage evaporative cooler will need to be exhausted directly (one-through flow) because the high humidity of the supply air. Few design solutions have been conceived to utilize the energy in the air like directing the exhaust air through two sheets of double glazed windows, thus reducing the solar energy absorbed through the glazing. Compared to energy required to achieve the equivalent cooling load with a compressor, single stage evaporative coolers consume less energy.

Passive direct evaporative cooling can occur anywhere that the evaporatively cooled water can cool a space without the assist of a fan. This can be achieved through use of fountains or more architectural designs such as the evaporative downdraft cooling tower, also called a "passive cooling tower". The passive cooling tower design allows outside air to flow in through the top of a tower that is constructed within or next to the building. The outside air comes in contact with water inside the tower either through a wetted membrane or a mister. As water evaporates in the outside air, the air becomes cooler and less buoyant and creates a downward flow in the tower. At the bottom of the tower, an outlet allows the cooler air into the interior. Similar to mechanical evaporative coolers, towers can be an attractive low-energy solution for hot and dry climate as they only require a water pump to raise water to the top of the tower. Energy savings from using a passive direct evaporating cooling strategy depends on the climate and heat load. For arid climates with a great wet-bulb depression, cooling towers can provide enough cooling during summer design conditions to be net zero. For example, a 371 m² (4,000 ft²) retail store in Tucson, Arizona with a sensible heat gain of 29.3 kJ/h (100,000 Btu/h) can be cooled entirely by two passive cooling towers providing 11890 m³/h (7,000 cfm) each.

For the Zion National Park visitors' center, which uses two passive cooling towers, the cooling energy intensity was 14.5 MJ/m² (1.28 kBtu/ft;), which was 77% less than a typical building in the western United States that uses 62.5 MJ/m² (5.5 kBtu/ft²). A study of field performance results in Kuwait revealed that power requirements for an evaporative cooler are approximately 75% less than the power requirements for a conventional packaged unit air-conditioner.

Indirect Evaporative Cooling

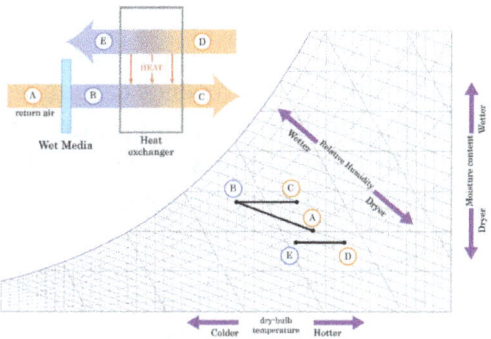

The process of indirect evaporative cooling

Indirect evaporative cooling (closed circuit) is a cooling process that uses direct evaporative cooling in addition to some type of heat exchanger to transfer the cool energy to the supply air. The cooled moist air from the direct evaporative cooling process never comes in direct contact with the conditioned supply air. The moist air stream is released outside or used to cool other external devices such as solar cells which are more efficient if kept cool. One indirect cooler manufacturer uses the so-called Maisotsenko cycle which employs an iterative (multi-step) heat exchanger that can reduce the temperature of product air to below the wet-bulb temperature, and can a approach the dew point. While no moisture is added to the incoming air the relative humidity (RH) does rise a little according to the Temperature-RH formula. Still, the relatively dry air resulting from indirect evaporative cooling allows inhabitants' perspiration to evaporate more easily, increasing the relative effectiveness of this technique. Indirect Cooling is an effective strategy for hot-humid climates that cannot afford to increase the moisture content of the supply air due to indoor air quality and human thermal comfort concerns. The following graphs describe the process of direct and indirect evaporative cooling with the changes in temperature, moisture content and relative humidity of the air.

Passive indirect evaporative cooling strategies are rare because this strategy involves an architectural element to act as a heat exchanger (for example a roof). This element can be sprayed with water and cooled through the evaporation of the water on this element. These strategies are rare due to the high use of water, which also introduces the risk of water intrusion and compromising building structure.

Two-stage Evaporative Cooling or Indirect-direct

In the first stage of a two-stage cooler, warm air is pre-cooled indirectly without adding humidity (by passing inside a heat exchanger that is cooled by evaporation on the outside). In the direct stage, the pre-cooled air passes through a water-soaked pad and picks up humidity as it cools. Since the air supply is pre-cooled in the first stage, less humidity is transferred in the direct stage, to reach the desired cooling temperatures. The result, according to manufacturers, is cooler air with a RH between 50-70%, depending on the climate, compared to a traditional system that produces about 70–80% relative humidity in the conditioned air.

In a *hybrid* design, direct or indirect cooling has been combined with vapor-compression or absorption air conditioning to increase the overall efficiency and to reduce the temperature below the wet-bulb limit.

Materials

Traditionally, evaporative *cooler pads* consist of excelsior (aspen wood fiber) inside a containment net, but more modern materials, such as some plastics and melamine paper, are entering use as cooler-pad media. Modern rigid media, commonly 8" or 12" thick, adds more moisture, and thus cools air more than typically much thinner aspen media. Another material which is sometimes used is corrugated cardboard.

Design Considerations

Water use

In arid and semi-arid climates, the scarcity of water makes water consumption a concern in cooling system design. From the installed water meters 420938 L (111,200 gal) of water were consumed during 2002 for the two passive cooling towers at the Zion National Park visitors' center. However, such concerns are addressed by experts who note that electricity generation usually requires a large amount of water, and evaporative coolers use far less electricity, and thus comparable water overall, and cost less overall, compared to chillers.

Shading

Allowing direct solar exposure to the media pads increases the evaporation rate. Sunlight may, however, degrade some media, in addition to heating up other elements of the evaporative cooling design. Therefore, shading is advisable in most applications.

Mechanical Systems

Apart from fans used in mechanical evaporative cooling, pumps are the only other piece of mechanical equipment required for the evaporative cooling process in both mechanical and passive applications. Pumps can be used for either recirculating the water to the wet media pad or providing water at very high pressure to a mister system for a passive cooling tower. Pump specifications will vary depending on evaporation rates and media pad area. The Zion National Park visitors' center uses a 250 W (1/3 HP) pump.

Exhaust

Exhaust ducts and open windows must be used at all times to allow air to continually escape the air-conditioned area. Otherwise, pressure develops and the fan or blower in the system is unable to push much air through the media and into the air-conditioned area. The evaporative system cannot function without exhausting the continuous supply of air from the air-conditioned area to the outside. By optimizing the placement of the cooled-air inlet, along with the layout of the house passages, related doors, and room windows, the system can be used most effectively to direct the cooled air to the required areas. A well-designed layout can effectively scavenge and expel the hot air from desired areas without the need for an above-ceiling ducted venting system. Continuous airflow is essential, so the exhaust windows or vents must not restrict the volume and passage of air being introduced by the evaporative cooling machine. One must also be mindful of the outside wind direction, as, for example, a strong hot southerly wind will slow or restrict the exhausted air from a south-facing window. It is always best to have the downwind windows open, while the upwind windows are closed.

Performance

Understanding evaporative cooling performance requires an understanding of psychrometrics. Evaporative cooling performance is variable due to changes in external temperature and humidity level. A residential cooler should be able to decrease the temperature of air to within 3 to 4° C (5 to 7° F) of the wet bulb temperature.

It is simple to predict cooler performance from standard weather report information. Because weather reports usually contain the dewpoint and relative humidity, but not the wet-bulb temperature, a psychrometric chart or a simple computer program must be used to compute the wet bulb temperature. Once the wet bulb temperature and the dry bulb temperature are identified, the cooling performance or leaving air temperature of the cooler may be determined.

For direct evaporative cooling, the direct saturation efficiency, ϵ, measures in what extent the temperature of the air leaving the direct evaporative cooler is close to the wet-bulb temperature of the entering air. The direct saturation efficiency can be determined as follows:

$$\epsilon = \frac{T_{e,db} - T_{l,db}}{T_{e,db} - T_{e,wb}}$$

Where:

ϵ = direct evaporative cooling saturation efficiency (%)

$T_{e,db}$ = entering air dry-bulb temperature (° C)

$T_{l,db}$ = leaving air dry-bulb temperature (° C)

$T_{e,wb}$ = entering air wet-bulb temperature (° C).

Evaporative media efficiency usually runs between 80% to 90%. Most efficient systems can lower the dry air temperature to 95% of the wet-bulb temperature, the least efficient systems only achieve 50%. The evaporation efficiency drops very little over time.

Typical aspen pads used in residential evaporative coolers offer around 85% efficiency while CELdek type of evaporative media offer efficiencies of >90% depending on air velocity. The CELdek media is more often used in large commercial and industrial installations.

As an example, in Las Vegas, with a typical summer design day of 42° C (108° F) dry bulb and 19° C (66° F) wet bulb temperature or about 8% relative humidity, the leaving air temperature of a residential cooler with 85% efficiency would be:

$$T_{l,db} = 42° \text{ C} - [(42° \text{ C} - 19° \text{ C}) \times 85\%] = 22.45° \text{ C or } 72.41° \text{ F}$$

However, either of two methods can be used to estimate performance:

- Use a psychrometric chart to calculate wet bulb temperature, and then add 5–7° F as described above.

- Use a rule of thumb which estimates that the wet bulb temperature is approximately equal to the ambient temperature, minus one third of the difference between the ambient temperature and the dew point. As before, add 5–7° F.

Some examples clarify this relationship:

- At 32° C (90° F) and 15% relative humidity, air may be cooled to nearly 16° C (61° F). The dew point for these conditions is 2° C (36° F).

- At 32° C and 50% relative humidity, air may be cooled to about 24° C (75° F). The dew point for these conditions is 20° C (68° F).

- At 40° C (104° F) and 15% relative humidity, air may be cooled to nearly 21° C (70° F). The dew point for these conditions is 8° C (46° F).

Because evaporative coolers perform best in dry conditions, they are widely used and most effective in arid, desert regions such as the southwestern USA and northern Mexico.

The same equation indicates why evaporative coolers are of limited use in highly humid environments: for example, a hot August day in Tokyo may be 30° C (86° F) with 85% relative humidity, 1,005 hPa pressure. This gives a dew point of 27.2° C (81.0° F) and a wet-bulb temperature of 27.88° C (82.18° F). According to the formula above, at 85% efficiency air may be cooled only down to 28.2° C (82.8° F) which makes it quite impractical.

Comparison to Air Conditioning

A misting fan

Comparison of evaporative cooling to refrigeration-based air conditioning:

Advantages

Less Expensive to Install and Operate

- Estimated cost for professional installation is about half or less that of central refrigerated air conditioning.

- Estimated cost of operation is 1/8 that of refrigerated air conditioning.

- No power spike when turned on due to lack of a compressor.

- Power consumption is limited to the fan and water pump, which have a relatively low current draw at start-up.

- The working fluid is water. No special refrigerants, such as ammonia or CFCs, are used that could be toxic, expensive to replace, contribute to ozone depletion and be subject to stringent licensing and environmental regulations.

Ease of Installation and Maintenance

- Equipment can be installed by mechanically-inclined users at drastically lower cost than refrigeration equipment which requires specialized skills and professional installation.

- The only two mechanical parts in most basic evaporative coolers are the fan motor and the water pump, both of which can be repaired or replaced at low cost and often by a mechanically inclined user, eliminating costly service calls to HVAC contractors.

Ventilation Air

- The frequent and high volumetric flow rate of air traveling through the building reduces the "age-of-air" in the building dramatically.

- Evaporative cooling increases humidity. In dry climates, this may improve comfort and decrease static electricity problems.

- The pad itself acts as a rather effective air filter when properly maintained; it is capable of removing a variety of contaminants in air, including urban ozone caused by pollution, regardless of very dry weather. Refrigeration-based cooling systems lose this ability whenever there is not enough humidity in the air to keep the evaporator wet while providing a frequent trickle of condensation that washes out dissolved impurities removed from the air.

Disadvantages

Performance

- Most evaporative coolers are unable to lower the air temperature as much as refrigerated air conditioning can.

- High dewpoint (humidity) conditions decrease the cooling capability of the evaporative cooler.

- No dehumidification. Traditional air conditioners remove moisture from the air, except in very dry locations where recirculation can lead to a buildup of humidity. Evaporative cooling adds moisture, and in humid climates, dryness may improve thermal comfort at higher temperatures.

Comfort

- The air supplied by the evaporative cooler is generally 80–90% relative humidity and can cause interior humidity levels as high as 65%; very humid air reduces the evaporation rate of moisture from the skin, nose, lungs, and eyes.

- High humidity in air accelerates corrosion, particularly in the presence of dust. This can considerably reduce the life of electronics and other equipment.

- High humidity in air may cause condensation of water. This can be a problem for some situations (e.g., electrical equipment, computers, paper, books, old wood).

- Odors and other outdoor contaminants may be blown into the building unless sufficient filtering is in place.

Water Use

- Evaporative coolers require a constant supply of water.

- Water high in mineral content (hard water) will leave mineral deposits on the pads and interior of the cooler. Depending on the type and concentration of minerals, possible safety hazards during the replacement and waste removal of the pads could be present. Bleed-off and refill (purge pump) systems can reduce but not eliminate this problem. Installation of an inline water filter (refrigerator drinking water/ice maker type) will drastically reduce the mineral deposits.

Maintenance Frequency

- Any mechanical components that can rust or corrode need regular cleaning or replacement due to the environment of high moisture and potentially heavy mineral deposits in areas with hard water.

- Evaporative media must be replaced on a regular basis to maintain cooling performance. Wood wool pads are inexpensive but require replacement every few months. Higher-efficiency rigid media is much more expensive but will last for a number of years proportional to the water hardness; in areas with very hard water, rigid media may only last for two years before mineral scale build-up unacceptably degrades performance.

- In areas with cold winters, evaporative coolers must be drained and winterized to protect the water line and cooler from freeze damage and then de-winterized prior to the cooling season.

Health Hazards

- An evaporative cooler is a common place for mosquito breeding. Numerous authorities consider an improperly maintained cooler to be a threat to public health.

- Mold and bacteria may be dispersed into interior air from improperly maintained or defective systems, causing sick building syndrome and adverse effects for asthma and allergy sufferers.

- Wood wool of dry cooler pads can catch fire even from small sparks.

Deep Water Source Cooling

Deep water cooling involves using naturally cold water as a heat sink in a heat exchange system, thereby eliminating the need for conventional air conditioning.

Conventional air conditioning functions by transferring heat from the air to a chilled medium, and then uses a compressor, motor, and refrigerant to transfer the heat from the chiller medium to the outdoors. If it is warmer outside than inside, heat must be pushed "uphill", a very energy intensive operation. Significant energy savings can be realized if heat can instead be transferred to a mass of cooler material with a high capacity for absorbing heat, such as water, eliminating the need for a compressor-based cooling cycle. Water is not only a good heat sink, it also has an unusual relation between its density and its temperature. Like most substances, water becomes denser as it cools, but unlike most substances it reaches a maximum density at 3.9 degrees Celsius. As a result, in winter, cold water on the surfaces of oceans and lakes cools and sinks through the warmer water below. In summer, the warm surface layers float on top of the cooler water below, as it is less dense. A layer of perpetually cold water is created below a certain depth, known as the hypolimnion.

Though the impact of deep water cooling is generally positive, some concerns have been raised that, if overused, the cold water source could experience "heat pollution", which would negatively affect habitat and species composition. In the oceans, such effects might occur at the local level, but the amount of heat involved is too small to have a large scale effect. Lakes are another matter.

Seawater Air Conditioning

Seawater Air Conditioning (SWAC) takes advantage of available deep cold water from the ocean, a river, or lake, to replace conventional AC systems. Such a system can also utilize cold lake or river water as the cold source.

SWAC feasibility studies for a variety of sites indicate that electrical consumption is typically reduced by 80 to 90 percent. Simple payback can be from three to seven years, and long term costs can be half that of a conventional air conditioning system.

Seawater Air Conditioning (SWAC) is an alternate-energy system that uses the cold water from the deep ocean (and in some cases a deep lake) to cool buildings. In some areas it is possible to reduce dramatically the power consumed by air conditioning (AC) systems; SWAC can be a cost-effective and attractive investment. It is an alternate energy for air conditioning.

Benefits of a SWAC System

The Seawater Air Condition Systems taps into a significant and highly valuable natural energy resource that is available at some coastal locations. The benefits of a seawater air conditioning system include:

- Large energy savings approaching 90%/.
- Proven technology.
- Short economic payback period.
- Environmentally friendly
- Costs are nearly independent of future energy price increases.
- No evaporative water consumption.
- Cold seawater availability for secondary applications.

References

- Needham, Joseph (1991). Science and Civilisation in China, Volume 4: Physics and Physical Technology, Part 2, Mechanical Engineering. Cambridge University Press. pp. 99, 151, 233. ISBN 978-0-521-05803-2.
- Window-air-conditioner: airconditioning-systems.com, Retrieved 15 July 2018
- Kingma, Boris; van Marken Lichtenbelt, Wouter (3 August 2015). "Energy consumption in buildings and female thermal demand". Nature Climate Change. doi:10.1038/NCLIMATE2741.
- How-the-window-air-conditioner-works-55241: brighthubengineering.com, Retrieved 11 April 2018
- "Negative Health Effects of Central AC". livestrong.com. Archived from the original on 28 January 2013. Retrieved 21 February 2013.

- Split-air-conditioner-system: networx.com, Retrieved 19 March 2018

- Nagengast, Bernard (February 1999). "A History of Comfort Cooling Using Ice" (PDF). ASHRAE Journal: 49. Retrieved 22 July 2013.

- Package-air-conditioner: airconditioning-systems.com, Retrieved 05 July 2018

- Shane Smith (2000). Greenhouse gardener's companion: growing food and flowers in your greenhouse or sunspace (2nd ed.). Fulcrum Publishing. p. 62. ISBN 978-1-55591-450-9.

- How-a-packaged-system-works, heating-cooling-101: goodmanmfg.com, Retrieved 30 June 2018

- Bonan, Gordon B. (13 June 2008). "Forests and Climate Change: Forcings, Feedbacks, and the Climate Benefits of Forests" (Submitted manuscript). Science. 320 (5882): 1444–9. Bibcode:2008Sci...320.1444B. doi:10.1126/science.1155121. PMID 18556546.

- Sea-water-air-conditioning: makai.com, Retrieved 22 May 2018

- "One in eight U.S. homes uses a programmed thermostat with a central air conditioning unit". U.S. Energy Information Administration. U.S. Department of Energy. 2017-07-19. Retrieved 2017-07-20.

5

Measurement for HVAC

Science and technology have undergone rapid developments in the past decade, which has resulted in innovation in HVAC systems. The following chapter elucidates a study of the measurements for HVAC and includes vital topics related to air flow meter, blower door, gas detector, thermostat, carbon dioxide sensor, etc.

Air Flow Meter

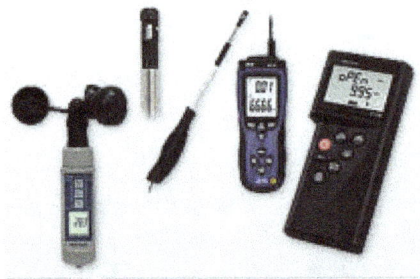

An airflow meter measures air velocity as well as air pressure. Depending on the type of application, the airflow meter is made as a hot-wire airflow meter, a vane airflow meter, a cup anemometer, a Pitot tube air flow meter or a pocket weather flow meter, all of which can measure air velocity as well as air pressure. Some can even detect the wind direction. The most common unit of measurement for air flow is m/s.

Airflow meters are basically mass flow meters, as they determine the air velocity and air pressure by measuring the mass flow of a medium, which is air in this case.

The results of measuring the air velocity can be stored in the airflow meter's memory, depending on the model. There are a lot of airflow meters available. The air flow meter is ideal for taking quick or constant measurements including monitoring of ventilation installations, process checking, industrial applications, use in private workshops or by nautical enthusiasts or for other hobbies, etc.

Two common types of air flow meters are used on automobiles, the hot wire and the vane meter.

Hot Wire

Hot wire sensors work by heating a thin wire suspended directly in an air intake system. The air flowing through the wire makes it cooler, and also lowers its electrical resistance. This variation of resistance corresponds to change in voltage signal which is sent back to the ECU.

Vane Meter

Vane meter is mounted by projecting a spring-loaded paddle directly into the engines air intake. Unlike the hot wire type, the air flowing through the vane moves the paddle proportionally. This movement causes for the voltage to vary, which is measured and sent back to the ECU.

Both type of air flow meters are sensitive electronic sensors. They need to be cleaned once in a while with electronic contact cleaner. Dirty air flow sensors will decrease your gas mileage and may shorten the life of your engine.

Use in Automobiles

An air flow meter is used in some automobiles to measure the quantity of air going into the internal combustion engine. All modern electronically controlled diesel engines use an air flow meter, as it is the only possible means of determining the air intake for them. In the case of a petrol engine, the electronic control unit (ECU) then calculates how much fuel is needed to inject into the cylinder ports. In the diesel engine, the ECU meters the fuel through the injectors into the engines cylinders during the compression stroke. An air flow meter is a special sensor which has been modified now by MEMS technology.

The vane (flap) type air flow meters (Bosch L-Jetronic and early Motronic EFI systems or Hitachi) actually measure air volume, whereas the later "hot wire" and "hot film" air mass meters measure the mass of air flow.

The flap-type meter includes a spring which returns the internal flap to the initial position. Sometimes if the spring is tensioned too tightly, it can restrict the incoming air and it would cause the intake air speed to increase when not opened fully.

Differential pressure is also used for air flow measurement purposes.

Failures

Air flow meters may fail or wear out. When this happens, engine performance will often decrease significantly, engine emissions will be greatly increased, and usually the Malfunction Indicator Lamp (MIL) will illuminate.

For R and D of Cars

In the development process of internal combustion engines with engine test stands, an air flow meter or air flow measuring unit is used for measuring the continuous gravimetric air consumption of combustion engines.

Industrial Environments

Air flow meters monitor air (compressed, forced, or ambient) in many manufacturing processes. In many industries, preheated air (called "combustion air") is added to boiler fuel just before fuel ignition to ensure the proper ratio of fuel to air for an efficient flame. Pharmaceutical factories and coal pulverizers use forced air as a means to force particle movement or ensure a dry atmosphere. Air flow is also monitored in mining and nuclear environments to ensure the safety of people.

Blower Door

A blower door is simply a diagnostic tool used to measure how much air filters out of your house (your home's "airtightness"). The blower door allows testers to apply a consistent and measurable pressure to the house so that houses can be compared accurately.

Working and Components of Blower Door

A blower door is a powerful fan that mounts into the frame of an exterior door. The fan pulls air out of the house, lowering the air pressure inside. The higher outside air pressure then flows in through all unsealed cracks and openings. The auditors may use a smoke pencil to detect air leaks. These tests determine the air infiltration rate of a building.

Blower doors consist of a frame and flexible panel that fit in a doorway, a variable-speed fan, a pressure gauge to measure the pressure differences inside and outside the home, and an airflow manometer and hoses for measuring airflow.

There are two types of blower doors: calibrated and uncalibrated. It is important that auditors use a calibrated door. This type of blower door has several gauges that measure the amount of air pulled out of the house by the fan. Uncalibrated blower doors can only locate leaks in homes. They provide no method for determining the overall tightness of a building. The calibrated blower door's data allow the auditor to quantify the amount of air leakage and the effectiveness of any air-sealing job.

Preparing for a Blower Door Test

Take the following steps to prepare your home for a blower door test:

- If you heat with wood, be sure all fires are completely out - not even coals - before the auditor arrives. Remove any ashes from open fireplaces.

- Plan to do a walk-through of your home with the auditor. Be prepared to point out areas that you know are drafty or difficult to condition comfortably.

- Expect the auditor to request access to all areas of your home including closets, built-in cabinets, attics, crawl spaces, and any unused rooms.

- The auditor will need to close all exterior doors and windows, open all interior doors, and close any fireplace dampers, doors, and woodstove air inlets.

- Expect the auditor to set controls on all atmospheric fossil fuel appliances to ensure that they do not fire during the test. The auditor should return them to the original position after the test.

- Expect the test to take up to an hour or more, depending on the complexity of your home.

Power Law Model of Airflow

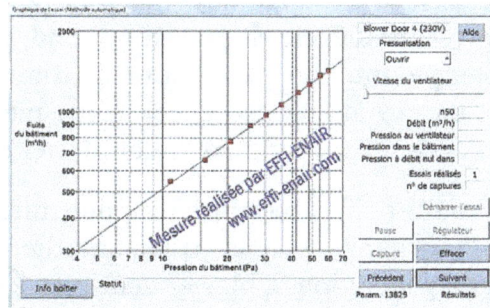

A typical graph of air leakage vs. pressure

Building leakage is described by a power law equation of flow through an orifice. The orifice flow equation is typically expressed as,

$$Q = C\Delta P^n$$

Q = Airflow (m³/s)

C = Air Leakage Coefficient

ΔP = Pressure Differential (Pa)

n = Pressure Exponent.

The C parameter reflects the size of the orifice, the ΔP is the pressure differential across the orifice, and the n parameter represents the characteristic shape of the orifice, with values ranging from 0.5 to 1, representing a perfect orifice and a very long, thin crack, respectively.

There are two airflows to be determined in blower door testing, airflow through the fan (Q_{Fan}) and airflow through the building envelope ($Q_{Building}$).

$$Q_{Fan} = C_{Fan}\Delta P_{Fan}^{n_{Fan}}$$

$$Q_{Building} = C_{Building}\Delta P_{Building}^{n_{Building}}$$

It is assumed in blower door analysis that mass is conserved, resulting in:

$$Q_{Fan} = Q_{Building}$$

Which results in:

$$C_{Fan}\Delta P_{Fan}^{n_{Fan}} = C_{Building}\Delta P_{Building}^{n_{Building}}$$

Fan airflow is determined using C_{Fan} and n_{Fan} values that are provided by the blower door manufacturer, and they are used to calculate Q_{Fan}. The multi-point blower door test procedure results in a series of known values of $Q_{n,\,Fan}$ and $\Delta P_{n,\,Building}$. Typical $\Delta P_{n,\,Building}$ values are ±5, 10, 20, 30, 40 and 50 pascal. Ordinary least squares regression analysis is then used to calculate the leakage characteristics of the building envelope: $C_{Building}$ and $n_{Building}$. These leakage characteristics of the building envelope can then be used to calculate how much airflow will be induced through the building envelope for a given pressure difference caused by wind, temperature difference or mechanical forces. 50 Pa can be plugged into the orifice-flow equation, along with the derived building C and n values to calculate airflow at 50 pascal. This same method can be used to calculate airflow at a variety of pressures, for use in creation of other blower door metrics.

An alternative approach to the multi-point procedure is to only measure fan airflow and building pressure differential at a single test point, such as 50 Pa, and then use an assumed pressure exponent, $n_{Building}$ in the analysis and generation of blower door metrics. This method is preferred by some for two main reasons: (1) measuring and recording one data point is easier than recording multiple test points, and (2) the measurements are least reliable at very low building pressure differentials, due both to fan calibration and to wind effects.

Air Density Corrections

In order to increase the accuracy of blower door test results, air density corrections should be applied to all airflow data. This must be done prior to the derivation of building air leakage coefficients ($C_{Building}$) and pressure exponents ($n_{Building}$). The following methods are used to correct blower door data to standard conditions.

For depressurization testing, the following equation should be used:

$$Q_{Corrected} = Q_{Measured} * \frac{\rho_{In}}{\rho_{Out}}$$

$Q_{Corrected}$ = Airflow corrected to actual air density

$Q_{Measured}$ = Airflow derived using C_{Fan} and n_{Fan}

ρ_{In} = Air density inside the building, during testing

ρ_{Out} = Air density outside the building, during testing.

For pressurization testing, the following equation should be used:

$$Q_{Corrected} = Q_{Measured} * \frac{\rho_{Out}}{\rho_{In}}$$

The values $\frac{\rho_{Out}}{\rho_{In}}$ and $\frac{\rho_{In}}{\rho_{Out}}$ are referred to as air density correction factors in product literature. They are often tabulated in easy to use tables in product literature, where a factor can be determined from outside and inside temperatures. If such tables are not used, the following equations will be required to calculate air densities.

ρ_{In} can be calculated in IP units using the following equation:

$$\rho_{In} = 0.07517 * (1 - \frac{0.0035666 * E}{528})^{5.2553} * (\frac{528}{T_{In} + 460})$$

ρ_{In} = Air density inside the building, during testing

E = Elevation above sea level (ft)

T_{In} = Indoor Temperature (F).

ρ_{Out} can be calculated in IP units using the following equation:

$$\rho_{Out} = 0.07517 * (1 - \frac{0.0035666 * E}{528})^{5.2553} * (\frac{528}{T_{Out} + 460})$$

ρ_{Out} = Air density outside the building, during testing

E = Elevation above sea level (ft)

T_{Out} = Outdoor Temperature (F).

In order to translate the airflow values derived using C_{Fan} and n_{Fan} from the blower door manufacturer to the actual volumetric airflow through the fan, use the following:

$$Q_{Actual} = Q_{Fan} * \sqrt{\frac{\rho_{Ref}}{\rho_{Actual}}}$$

Q_{Actual} = Actual volumetric airflow through the fan

Q_{Fan} = Volumetric airflow calculated using manufacturer's coefficients or software

ρ_{Ref} = Reference air density (typically 1.204 for kg/m³ or 0.075 for lb/ft³)

ρ_{Actual} = Actual density of air going through the fan ρ_{In} for depressurization and ρ_{Out} for pressurization.

Blower Door Metrics

Blower door installation

Depending on how a blower door test is performed, a wide variety of airtightness and building airflow metrics can be derived from the gathered data. Some of the most common metrics and their variations are discussed below. The examples below use the SI pressure measurement unit Pascal (pa). Imperial measurement units are commonly water column inches (WC Inch or IWC). The conversion rate is 1 WC inch = 249 Pa. Examples below use the commonly accepted pressure of 50pa which is 20% of 1 IWC.

Airflow at a Specified Building Pressure

This is the first metric that results from a Blower Door Test. The airflow, (Imperial in Cubic Feet/minute; SI in liters/second) at a given building-to-outside pressure differential, 50 pascal (Q_{50}). This standardized single-point test allows for comparison between homes measured at the same reference pressure. This is a raw number reflecting only the flow of air through the fan. Homes of different sizes and similar envelope quality will have different results in this test.

Airflow per Unit Surface Area or Floor Area

Often, an effort is made to control for building size and layout by normalizing the airflow at a specified building pressure to either the building's floor area or to its total surface area. These values are generated by taking the airflow rate through the fan and dividing by the area. These metrics are most used to assess construction and building envelope quality, because they normalize the total building leakage area to the total amount of area through which that leakage could occur. In other words, how much leakage occurs per unit area of wall, floor, ceiling, etc.

Air Changes per Hour at a Specified Building Pressure

Another common metric is the air changes per hour at a specified building pressure, again, typically at 50 Pa (ACH_{50}).

$$ACH_{50} = \frac{Q_{50} * 60}{V_{Building}}$$

ACH_{50} = Air changes per hour at 50 pascal (h^{-1})

Q_{50} = Airflow at 50 pascal (ft^3/minute)

$V_{Building}$ = Building volume (ft^3)

This normalizes the airflow at a specified building pressure by the building's volume, which allows for more direct comparison of homes of different sizes and layouts. This metric indicates the rate at which the air in a building is replaced with outside air, and as a result, is an important metric in determinations of indoor air quality.

Effective Leakage Area

In order to take values generated by fan pressurization and to use them in determining natural air exchange, the effective leakage area of a building must be calculated. Each gap and crack in the building envelope contributes a certain amount of area to the total leakage area of the building. The Effective Leakage Area assumes that all of the individual leakage areas in the building are combined into a single idealized orifice or hole. This value is typically described to building owners as the area of a window that is open 24/7, 365 in their building. The ELA will change depending on the reference pressure used to calculate it. 4 Pa is typically used in the US, whereas a reference pressure of 10 Pa is used in Canada. It is calculated as follows:

$$ELA = C_{Building} * \sqrt{\frac{\rho}{2}} * \Delta P_{Ref}^{n_{Building}-0.5}$$

ELA = Effective Leakage Area (m² or in²)

$C_{Building}$ = Building air leakage coefficient

ρ = Air density (kg/m³ or lb/in³), typically a standard density is used

ΔP_{Ref} = Reference Pressure (Pa or lb_{Force}/in²), typically 4 Pa in US and 10 Pa in Canada

$n_{Building}$ = Building pressure exponent.

It is essential that units are carefully conserved in these calculations. $C_{Building}$ and $n_{Building}$ should be calculated using SI units, and ρ and $\Delta P_{Reference}$ should be kg/m³ and pascal, respectively. Alternatively, $C_{Building}$ and $n_{Building}$ can be calculated using Imperial units, with ρ and $\Delta P_{Reference}$ being lb/ft³ and lb_{Force}/in², respectively.

The ELA can be used, along with the Specific Infiltration Rate (s) derived using the LBNL infiltration model, to determine airflow rate through the building envelope throughout the year.

Leakage Area per Unit Floor or Surface Area

Leakage area estimates can also be normalized for the size of the enclosure being tested, For example, the LEED Green Building Rating System has set an airtightness standard for multifamily dwelling units of 1.25 square inches (8.1 cm²) of leakage area per 100 square feet (9.3 m²) of enclosure area, to control tobacco smoke between units. This is equal to 0.868 cm²/m².

Normalized Leakage

Normalized leakage is a measure of the tightness of a building envelope relative to the building size and number of stories. Normalized Leakage is defined in ASHRAE Standard 119 as:

$$NL = 1000 * (\frac{ELA}{A_{Floor}}) * (\frac{H}{H_{Ref}})^{0.3}$$

NL = Normalized leakage

ELA = Effective Leakage Area (m² or in²)

A_{Floor} = Building floor area (m² or in²)

H = Building Height (m or in)

H_{Ref} = Reference height (2.5 metres (98 in)).

Applications

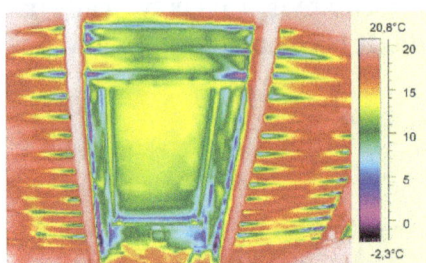

Infrared view of leaky window pressurized by blower door testing

Visible light view of window under test

Blower doors can be used in a variety of types of testing. These include (but are not limited to):

- Testing residential and commercial buildings for air tightness.

- Testing buildings at mid-construction to identify and correct any failures in the enclosure.

- Testing buildings for compliance with standards for energy efficiency, such as the IECC and ASHRAE.

- Testing building envelopes and window frames for water tightness and rain penetration.

- NFPA Clean Agent Retention testing (this type of testing is usually described as a *door fan test* rather than a blower door test).

- Duct leakage testing of forced air heating/cooling systems - both supply (vents) ducts and return ducts can be tested to determine if and how much they leak air. A duct test can be combined with a blower door test to measure the total leakage to outside, measuring effective leakage to the outside of the house only.

- Finding air leaks in a building using an infrared camera while the house is depressurized. A blower door is not mandatory for an infrared reading, but the drawing in of outside air temperatures exaggerates temperature changes and facilitates the spotting of envelope leaks.

Gas Detector

Gas detectors measure and indicate the concentration of certain gases in an air via different technologies. Typically employed to prevent toxic exposure and fire, gas detectors are often battery operated devices used for safety purposes. They are manufactured as portable or stationary (fixed) units and work by signifying high levels of gases through a series of audible or visible indicators, such as alarms, lights or a combination of signals. While many of the older, standard gas detector units were originally fabricated to detect one gas, modern multifunctional or multi-gas devices are capable of detecting several gases at once. Some detectors may be utilized as individual units to monitor small workspace areas, or units can be combined or linked together to create a protection system.

As detectors measure a specified gas concentration, the sensor response serves as the reference point or scale. When the sensors response surpasses a certain pre-set level, an alarm will activate to warn the user. There are various types of detectors available and the majority serves the same function: to monitor and warn of a dangerous gas level. However, when considering what type of detector to install, it is helpful to consider the different sensor technologies.

Gas Detector Technologies

Gas detectors are categorized by the type of gas they detect: combustible or toxic. Within this broad categorization, they are further defined by the technology they use: catalytic and infrared sensors detect combustible gases and electrochemical and metal oxide semiconductor technologies generally detect toxic gases.

Measurement of Toxic Gases

Electrochemical sensors or cells are most commonly used in the detection of toxic gases like carbon monoxide, chlorine and nitrogen oxides. They function via electrodes signals when a gas is detected. Generally, these types of detectors are highly sensitive and give off warning signals via electrical currents. Various manufacturers produce these detectors with a digital display.

Metal Oxide Semiconductors, or MOS, are also used for detecting toxic gases (commonly carbon monoxide) and work via a gas sensitive film that is composed of tin or tungsten oxides. The sensitive film reacts with gases, triggering the device when toxic levels are present. Generally, metal oxide sensors are considered efficient due their ability to operate in low-humidity ranges. In addition, they are able to detect a range of gases, including combustibles.

Measurement of Combustible Gases

Catalytic sensors represent a large number of gas detector devices that are manufactured today. This technology is used to detect combustible gases such as hydrocarbon, and works via catalytic oxidation. The sensors of this type of detector are typically constructed from a platinum treated wire coil. As a combustible gas comes into contact with the catalytic surface, it is oxidized and the wiring resistance is changed by heat that is released. A bridge circuit is typically used to indicate the resistance change.

Infrared sensors or IR detectors work via a system of transmitters and receivers to detect combustible gases, specifically hydrocarbon vapors. Typically, the transmitters are light sources and receivers are light detectors. If a gas is present in the optical path, it will interfere with the power of the light transmission between the transmitter and receiver. The altered state of light determines if and what type of gas is present.

Common Gas Detector Applications

Although detectors are an essential application for home and commercial safety, they are also employed in numerous industrial industries. Gas detectors are used in welding shops to detect combustibles and toxics and in nuclear plants, to detect combustibles. They are also commonly used to detect hazardous vapors in wastewater treatment plants.

Gas detectors are very efficient in confined spaces where there is no continuous employee occupancy. Such spaces include tanks, pits, vessels and storage bins. Detectors may also be placed at a site to detect toxins prior to occupant entry.

Types

Gas detectors can be classified according to the operation mechanism (semiconductors, oxidation, catalytic, photoionization, infrared, etc.). Gas detectors come packaged into two main form factors: portable devices and fixed gas detectors.

Portable detectors are used to monitor the atmosphere around personnel and are either hand-held or worn on clothing or on a belt/harness. These gas detectors are usually battery operated. They transmit warnings via audible and visible signals, such as alarms and flashing lights, when dangerous levels of gas vapors are detected.

Fixed type gas detectors may be used for detection of one or more gas types. Fixed type detectors are generally mounted near the process area of a plant or control room, or an area to be protected, such as a residential bedroom. Generally, industrial sensors are installed on fixed type mild steel structures and a cable connects the detectors to a SCADA system for continuous monitoring. A tripping interlock can be activated for an emergency situation.

Electrochemical

Electrochemical gas detectors work by allowing gases to diffuse through a porous membrane to an electrode where it is either chemically oxidized or reduced. The amount of current produced is determined by how much of the gas is oxidized at the electrode, indicating the concentration of the gas. Manufactures can customize electrochemical gas detectors by changing the porous barrier to allow for the detection of a certain gas concentration range. Also, since the diffusion barrier is a physical/mechanical barrier, the detector tended to be more stable and reliable over the sensor's duration and thus required less maintenance than other early detector technologies.

However, the sensors are subject to corrosive elements or chemical contamination and may last only 1–2 years before a replacement is required. Electrochemical gas detectors are used in a wide variety of environments such as refineries, gas turbines, chemical plants, underground gas storage facilities, and more.

Catalytic Bead (Pellistor)

Catalytic bead sensors are commonly used to measure combustible gases that present an explosion hazard when concentrations are between the lower explosion limit (LEL) and upper explosion limit

(UEL). Active and reference beads containing platinum wire coils are situated on opposite arms of a Wheatstone bridge circuit and electrically heated, up to a few hundred degrees C. The active bead contains a catalyst that allows combustible compounds to oxidize, thereby heating the bead even further and changing its electrical resistance. The resulting voltage difference between the active and passive beads is proportional to the concentration of all combustible gases and vapors present. The sampled gas enters the sensor through a sintered metal frit, which provides a barrier to prevent an explosion when the instrument is carried into an atmosphere containing combustible gases. Pellistors measure essentially all combustible gases, but they are more sensitive to smaller molecules that diffuse through the sinter more quickly. The measureable concentration ranges are typically from a few hundred ppm to a few volume percent. Such sensors are inexpensive and robust, but require a minimum of a few percent oxygen in the atmosphere to be tested and they can be poisoned or inhibited by compounds such as silicones, mineral acids, chlorinated organic compounds, and sulfur compounds.

Photoionization

Photoionization detectors (PIDs) use a high-photon-energy UV lamp to ionize chemicals in the sampled gas. If the compound has an ionization energy below that of the lamp photons, an electron will be ejected, and the resulting current is proportional to the concentration of the compound. Common lamp photon energies include 10.0 eV, 10.6 eV and 11.7 eV; the standard 10.6 eV lamp lasts for years, while the 11.7 eV lamp typically last only a few months and is used only when no other option is available. A broad range of compounds can be detected at levels ranging from a few ppb to several thousand ppm. Detectable compound classes in order of decreasing sensitivity include: aromatics and alkyl iodides; olefins, sulfur compounds, amines, ketones, ethers, alkyl bromides and silicate esters; organic esters, alcohols, aldehydes and alkanes; H_2S, NH_3, PH_3 and organic acids. There is no response to standard components of air or to mineral acids. Major advantages of PIDs are their excellent sensitivity and simplicity of use; the main limitation is that measurements are not compound-specific. Recently PIDs with pre-filter tubes have been introduced that enhance the specificity for such compounds as benzene or butadiene. Fixed, hand-held and miniature clothing-clipped PIDs are widely used for industrial hygiene, hazmat, and environmental monitoring.

Infrared Point

Infrared (IR) point sensors use radiation passing through a known volume of gas; energy from the sensor beam is absorbed at certain wavelengths, depending on the properties of the specific gas. For example, carbon monoxide absorbs wavelengths of about 4.2-4.5 μm. The energy in this wavelength is compared to a wavelength outside of the absorption range; the difference in energy between these two wavelengths is proportional to the concentration of gas present.

This type of sensor is advantageous because it does not have to be placed into the gas to detect it and can be used for remote sensing. Infrared point sensors can be used to detect hydrocarbons and other infrared active gases such as water vapor and carbon dioxide. IR sensors are commonly found in waste-water treatment facilities, refineries, gas turbines, chemical plants, and other facilities where flammable gases are present and the possibility of an explosion exists. The remote sensing capability allows large volumes of space to be monitored.

Engine emissions are another area where IR sensors are being researched. The sensor would detect high levels of carbon monoxide or other abnormal gases in vehicle exhaust and even be integrated with vehicle electronic systems to notify drivers.

Infrared Imaging

Infrared image sensors include active and passive systems. For active sensing, IR imaging sensors typically scan a laser across the field of view of a scene and look for backscattered light at the absorption line wavelength of a specific target gas. Passive IR imaging sensors measure spectral changes at each pixel in an image and look for specific spectral signatures that indicate the presence of target gases. The types of compounds that can be imaged are the same as those that can be detected with infrared point detectors, but the images may be helpful in identifying the source of a gas.

Semiconductor

Semiconductor sensors detect gases by a chemical reaction that takes place when the gas comes in direct contact with the sensor. Tin dioxide is the most common material used in semiconductor sensors, and the electrical resistance in the sensor is decreased when it comes in contact with the monitored gas. The resistance of the tin dioxide is typically around 50 kΩ in air but can drop to around 3.5 kΩ in the presence of 1% methane. This change in resistance is used to calculate the gas concentration. Semiconductor sensors are commonly used to detect hydrogen, oxygen, alcohol vapor, and harmful gases such as carbon monoxide. One of the most common uses for semiconductor sensors is in carbon monoxide sensors. They are also used in breathalyzers. Because the sensor must come in contact with the gas to detect it, semiconductor sensors work over a smaller distance than infrared point or ultrasonic detectors.

Ultrasonic

Ultrasonic gas leak detectors are not gas detectors per se. They detect the acoustic emission created when a pressured gas expands in a low pressure area through a small orifice (the leak). They use acoustic sensors to detect changes in the background noise of its environment. Since most high-pressure gas leaks generate sound in the ultrasonic range of 25 kHz to 10 MHz, the sensors are able to easily distinguish these frequencies from background acoustic noise which occurs in the audible range of 20 Hz to 20 kHz. The ultrasonic gas leak detector then produces an alarm when there is an ultrasonic deviation from the normal condition of background noise. Ultrasonic gas leak detectors cannot measure gas concentration, but the device is able to determine the leak rate of an escaping gas because the ultrasonic sound level depends on the gas pressure and size of the leak.

Ultrasonic gas detectors are mainly used for remote sensing in outdoor environments where weather conditions can easily dissipate escaping gas before allowing it to reach leak detectors that require contact with the gas to detect it and sound an alarm. These detectors are commonly found on offshore and onshore oil/gas platforms, gas compressor and metering stations, gas turbine power plants, and other facilities that house a lot of outdoor pipeline.

Holographic

Holographic gas sensors use light reflection to detect changes in a polymer film matrix containing a hologram. Since holograms reflect light at certain wavelengths, a change in their composition can generate a colorful reflection indicating the presence of a gas molecule. However, holographic sensors require illumination sources such as white light or lasers, and an observer or CCD detector.

Calibration

All gas detectors must be calibrated on a schedule. Of the two form factors of gas detectors, portables must be calibrated more frequently due to the regular changes in environment they experience. A typical calibration schedule for a fixed system may be quarterly, bi-annually or even annually with more robust units. A typical calibration schedule for a portable gas detector is a daily "bump test" accompanied by a monthly calibration. Almost every portable gas detector requires a specific calibration gas which is available from the manufacturer. In the US, the Occupational Safety and Health Administration (OSHA) may set minimum standards for periodic recalibration.

Challenge (Bump) Test

Because a gas detector is used for employee/worker safety, it is very important to make sure it is operating to manufacturer's specifications. Australian standards specify that a person operating any gas detector is strongly advised to check the gas detector's performance each day and that it is maintained and used in accordance with the manufacturers instructions and warnings.

A challenge test should consist of exposing the gas detector to a known concentration of gas to ensure that the gas detector will respond and that the audible and visual alarms activate. It is also important inspect the gas detector for any accidental or deliberate damage by checking that the housing and screws are intact to prevent any liquid ingress and that the filter is clean, all of which can affect the functionality of the gas detector. The basic calibration or challenge test kit will consist of calibration gas/regulator/calibration cap and hose (generally supplied with the gas detector) and a case for storage and transport. Because 1 in every 2,500 untested instruments will fail to respond to a dangerous concentration of gas, many large businesses use an automated test/calibration station for bump tests and calibrate their gas detectors daily.

Oxygen Concentration

Oxygen deficiency gas monitors are used for employee and workforce safety. Cryogenic substances such as liquid nitrogen (LN_2), liquid helium (He), and liquid argon (Ar) are inert and can displace oxygen (O_2) in a confined space if a leak is present. A rapid decrease of oxygen can provide a very dangerous environment for employees, who may not notice this problem before they suddenly lose consciousness. With this in mind, an oxygen gas monitor is important to have when cryogenics are present. Laboratories, MRI rooms, pharmaceutical, semiconductor, and cryogenic suppliers are typical users of oxygen monitors.

Oxygen fraction in a breathing gas is measured by electro-galvanic oxygen sensors. They may be used stand-alone, for example to determine the proportion of oxygen in a nitrox mixture used in scuba diving, or as part of feedback loop which maintains a constant partial pressure of oxygen in a rebreather.

Hydrocarbons and VOCs

Detection of hydrocarbons can be based on the mixing properties of gaseous hydrocarbons – or other volatile organic compounds (VOCs) – and the sensing material incorporated in the sensor. The selectivity and sensitivity depends on the molecular structure of the VOC and the concentration; however, it is difficult to design a selective sensor for a single VOC. Many VOC sensors detect using a fuel-cell method.

VOCs in the environment or certain atmospheres can be detected based on different principles and interactions between the organic compounds and the sensor components. There are electronic devices that can detect ppm concentrations despite not being particularly selective. Others can predict with reasonable accuracy the molecular structure of the volatile organic compounds in the environment or enclosed atmospheres and could be used as accurate monitors of the chemical fingerprint and further as health monitoring devices.

Solid-phase microextraction (SPME) techniques are used to collect VOCs at low concentrations for analysis.

Direct injection mass spectrometry techniques are frequently utilized for the rapid detection and accurate quantification of VOCs. PTR-MS is among the methods that have been used most extensively for the on-line analysis of biogenic and antropogenic VOCs. Recent PTR-MS instruments based on time-of-flight mass spectrometry have been reported to reach detection limits of 20 pptv after 100 ms and 750 ppqv after 1 min measurement (signal integration) time. The mass resolution of these devices is between 7000 and 10,500 $m/\Delta m$, thus it is possible to separate most common isobaric VOCs and quantify them independently.

Considerations for Detection of Hydrocarbon Gases/risk Control

- Methane is lighter than air (possibility of accumulation under roofs).
- Ethane is slightly heavier than air (possibility of pooling at ground levels/pits).
- Propane is heavier than air (possibility of pooling at ground levels/pits).
- Butane is heavier than air (possibility of pooling at ground levels/pits).

Ammonia

Gaseous ammonia is continuously monitored in industrial refrigeration processes and biological degradation processes, including exhaled breath. Depending on the required sensitivity, different types of sensors are used (e.g., flame ionization detector, semiconductor, electrochemical, photonic membranes). Detectors usually operate near the lower exposure limit of 25ppm; however, ammonia detection for industrial safety requires continuous monitoring above the fatal exposure limit of 0.1%.

Carbon Dioxide Sensor

A carbon dioxide sensor measures gaseous carbon dioxide levels by monitoring the amount of infrared (IR) radiation absorbed by carbon dioxide molecules. Measuring carbon dioxide is critical in monitoring many industrial processes and indoor air quality.

Types of Carbon Dioxide Sensor

The following are two major types of carbon dioxide sensors:

Nondispersive Infrared (NDIR) Carbon Dioxide Sensors

NDIR sensors are spectroscopic sensors that detect carbon dioxide in a gaseous environment by its characteristic absorption. It includes an infrared detector, an interference filter, a light tube and an infrared source. The gas is either pumped or diffused into the light tube, and the electronics measure the absorption of the wavelength of light.

Chemical Carbon Dioxide Sensors

Chemical carbon dioxide gas sensors consist of sensitive layers based on polymer- or heteropolysiloxane. They have very low energy consumption and can be reduced in size to fit into microelectronic-based systems.

Working Principle

The carbon dioxide gas sensor measures gaseous carbon dioxide levels by detecting the quantity of IR radiation absorbed by carbon dioxide molecules. The sensor employs a hot metal filament that acts as an IR source to generate IR radiation.

The IR source is located at one end of a sensor tube, and another end is provided with an IR sensor. The IR sensor measures the amount of radiation that passes through the sample without being absorbed by the carbon dioxide molecules.

The sensor measures the IR radiation absorbed in the narrow band at 4260 nm. The greater the absorbing gas concentration in the sampling tube, the lesser the amount of radiation from the source.

As a result of this increase in temperature, a voltage is generated, amplified and read by an interface system. Meanwhile, the carbon dioxide gas diffuses via the sensor tube by the eight vent holes.

Applications

Some of the major applications of carbon dioxide sensors include the following:

- It can be used for HVAC applications to monitor the quality of air.
- It is used to monitor fermentation, aerobic respiration, photosynthesis and other carbon dioxide consuming or producing processes.

Clean Air Delivery Rate

The Clean Air Delivery Rate, or CADR for short, of an air purifier is a numerical value assigned to an air purifier based on independent tests performed for the Association of Home Appliance Manufacturers (or AHAM). AHAM has been measuring the CADRs of air cleaners since the 1980s.

The CADR definition is *an indication of the volume of filtered air delivered by a portable air purifier*. It goes on to say that CADR describes how well the cleaner reduces tobacco smoke, pollen and dust. In other words, the CADR is an attempt to provide a uniform, objective standard by which potential buyers can easily evaluate the effectiveness of an air cleaner.

Given the potentially bewildering variety of air purifiers available in the marketplace, such a standard is highly desirable. AHAM's own literature on CADR also emphasizes that CADR ratings are valid "regardless of the particle removal technology utilized," so it does not matter whether the purifier uses, for example, a HEPA filter or an ionizer. CADR measures only the end results.

CADR Determination

An air cleaner is given a CADR through a relatively easy to understand process, which is called the ANSI/AHAM AC-1 standard which measures CADR requirements:

- The purifier is placed in a testing chamber of 1008 cubic feet.

- Before the purifier is activated, the amount of contaminants in the room is measured.

- The purifier is activated for a period of twenty minutes, during which time the amount of contaminants is periodically re-evaluated.

- Finally, the reduction in contaminants is compared to their natural rate of decay.

When this test is over, testers proceed to give the unit its rating.

AHAM's Three CADR Ratings

There are three CADR ratings given on the AHAM seal. These represent the air cleaners' effectiveness against three different common indoor air pollutants: tobacco smoke, pollen, and dust. AHAM recommends a "two thirds" rule when it comes to the first rating: "You'll always want a unit with a tobacco smoke CADR at least 2/3 your room's area."

Highest Possible CADR Ratings

Due to the fact that the tests always use the same room size, there is an upper limit to meaningful CADR ratings:

- Tobacco Smoke 10 to 450 CADR

- Dust 10 to 400 CADR

- Pollen 25 to 450 CADR.

Uses of CADR Ratings

CADR testing has a history of nearly 30 years, and in that time, its use has spread. Reputable organizations beyond the air purification industry have adopted CADR, including the Environmental Protection Agency, the Federal Trade Commission, and the independent watchdog organization Consumers Union.

There is no reason to believe that CADRs are not effective, within limits. Nonetheless, as always, a certain measure of caution is advisable.

Limits of CADR Ratings

Effectiveness of Small Particle Filtration

CADR numbers do not adequately inform customers of a purifier's ability to filter out very small particles. Yet these are the most numerous, and often the most dangerous, pollutants. Looking at an air cleaner's Clean Air Delivery Rate, you cannot tell how effective it is in dealing specifically with these tiniest bits of matter. Many air purifiers eliminate larger particles very well but are almost completely incapable of removing smaller ones from circulation.

Measurement of Gas Filtration

The test does absolutely nothing to measure a purifier's ability to rid the air of gaseous pollutants, such as dangerous Volatile Organic Compounds. The three CADR numbers all refer to particulate pollution only, which is very different from gaseous pollution. Air purifiers vary wildly in their ability to deal with the latter. Because cleaning gas pollution requires different mechanisms from cleaning solid particle pollution, a purifier's effectiveness at reducing gas is completely unrelated to its effectiveness at reducing particles.

Effectiveness of Decline Over Time

Because the test is so short, it does not account for the fact that some purifiers will decline in effectiveness over time.

Purifiers Run at their Highest Settings

Fourth, in the ANSI/AHAM AC-1 test, purifiers are always run on their highest setting. But users will often use lower settings to reduce noise, and this can drastically reduce the efficiency of an air cleaner.

Minimum Efficiency Reporting Value

MERV is the Minimum Efficiency Reporting Value, which is the scale designed by the American Society of Heating, Refrigerating, and Air Conditioning Engineers (ASHRAE) to rate the effectiveness of air filters. Basically, the higher MERV rating indicates higher filtering performance.

The most recommended MERV ratings for residential filters are between MERV 7 and MERV 13. In fact, filters with these MERV ratings perform as efficiently as true HEPA filters. They all basically trap the same range of particulates, particulates such as dust particles, mold spores, dust mite debris, tobacco smoke, pet dander, and pollen, but each filter has a higher capturing capacity the higher the MERV.

There are really two different components of a MERV rating. One is the ability to remove large airborne particles, or particles that are between 1 and 10 microns. These "large" particles include pet dander, pollen grains, dust mite debris, etc. However, the bulk of indoor air pollutants are in a smaller size range. MERV also rates the filter's ability to remove submicron particles, those that are smaller than 1 micron; this rating is called the Microparticle Performance Rating (MPR), and some filters are given just this rating.

Chart below for filter MERV descriptions:

Merv Value	The filter will trap Average Particle Size Efficiency 0.3 - 1.0 Micron	The filter will trap Average Particle Size Efficiency 1.0 - 3.0 Micron	The filter will trap Average Particle Size Efficiency 3.0 - 10.0 Micron	Types of things these filters will trap
MERV 1	-	-	Less than 20%	Pollen, dust mites, standing dust, spray paint dust, carpet fibers
MERV 2	-	-	Less than 20%	
MERV 3	-	-	Less than 20%	
MERV 4	-	-	Less than 20%	
MERV 5	-	-	20% - 34%	Mold spores, hair spray, fabric protector, cement dust
MERV 6	-	-	35% - 49%	
MERV 7	-	-	50% - 69%	
MERV 8	-	-	70% - 85%	
MERV 9	-	Less than 50%	85% or better	Humidifier dust, lead dust, auto emissions, milled flour
MERV 10	-	50% - 64%	85% or better	
MERV 11	-	65% - 79%	85% or better	
MERV 12	-	80% - 89%	90% or better	
MERV 13	Less than 75%	90% or better	90% or better	Bacteria, most tobacco smoke, proplet Nuceli (sneeze)
MERV 14	75% - 84%	90% or better	90% or better	
MERV 15	85% - 94%	90% or better	90% or better	
MERV 16	95% or better	90% or better	90% or better	

Details about the Most Common Filters for Residential Use

MERV 6

MERV 6 is the lowest MERV rating offered by AirFilters.com. Though not recommended for allergy and asthma sufferers, filters with a MERV 6 rating are 8 times more efficient than the regular cheap fiberglass filters. Filters with a MERV 6 rating will remove large particles like allergens, dirt, debris, lint, dust, etc. The advantage of having a MERV 6 rated filter is that it is more affordable.

MERV 8

If you are looking for an ordinary household air filter for maintaining efficient airflow and better air system performance, then MERV 8 is the filter you need. MERV 8 rated filters are high quality filters. MERV 8 filters remove basically the same particulates as a MERV 6, but they are more efficient; with a dust spot efficiency of 30-35%, they are 20 times more efficient than regular fiberglass filters.

MERV 11

Filters with a MERV 11 rating have a dust spot efficiency of 60-65% and are 30 times more efficient than regular fiberglass filters. MERV 11 rated filters are good filters for households with breathing sufferers. Not only do they provide superior residential filtration, but they also provide a healthier breathing environment.

MERV 13

If you suffer from breathing problems such as allergies or asthma, then AirFilters.com recommends a filter with a MERV 13 rating. MERV 13 rated filters have a dust spot efficiency of 80-90%, and they are highly efficient filters both for residential and commercial use. MERV 13 filters remove a higher range of particulates, including larger amounts of bacteria and even some viruses, and they are best for people suffering from breathing problems. Also, for their high capturing capacity, they still maintain a low resistance to airflow.

Thermostat

Thermostat is a device to detect temperature changes for the purpose of maintaining the temperature of an enclosed area essentially constant. In a system including relays, valves, switches, etc., the thermostat generates signals, usually electrical, when the temperature exceeds or falls below the desired value. It usually is used to control the flow of fuel to a burner, of electric current to a heating or cooling unit, or of a heated or cooled gas or liquid into the area it serves. The thermostat is also an element in some types of fire-detection warning systems.

Working of Mechanical Thermostat

Mechanical thermostats use this idea of thermal expansion to switch an electric circuit on and and off. The two most common types use bimetallic strips and gas-filled bellows.

Bimetallic Strips

A traditional thermostat has two pieces of different metals bolted together to form what's called a bimetallic strip (or bimetal strip). The strip works as a bridge in an electrical circuit connected to your heating system. Normally the "bridge is down", the strip carries electricity through the circuit, and the heating is on. When the strip gets hot, one of the metals expands more than the other so the whole strip bends very slightly. Eventually, it bends so much that it breaks open the circuit. The "bridge is up", the electricity instantly switches off, the heating cuts out, and the room starts to cool.

But then what happens? As the room cools, the strip cools too and bends back to its original shape. Sooner or later, it snaps back into the circuit and makes the electricity flow again, so the heating switches back on. By adjusting the temperature dial, you change the temperature at which the circuit switches on and off. Because it takes some time for the metal strip to expand and contract, the heating isn't constantly switching on and off every few seconds, which would be pointless (and quite irritating); depending on how well-insulated your home is, and how cold it is outside, it might take an hour or more for the thermostat to switch back on once it's switched off.

Gas-filled Bellows

The trouble with bimetallic strips is that they take a long time to heat up or cool down, so they don't react quickly to temperature changes. An alternative design of thermostat senses temperature changes more quickly using a pair of metal discs with a gas-filled bellows in between. The discs have a large surface area so they react quickly to heat and they're corrugated (they have ridges in them) to make them springy and flexible. When the room warms up, the gas in the bellows expands and forces the discs apart. The inner disc pushes against a microswitch in the middle of the thermostat turning the electric circuit (and the heating) off. As the room cools, the gas in the bellows contracts and the metal discs are forced back together. The inner disc moves away from the microswitch, switching on the electric circuit and turning the heating on again. You can also find corrugated bellows thermostats in other applications (for example, older cars), and, instead of gas, they're sometimes filled with a volatile (low-boiling) liquid such as a diluted alcohol; the exact chemical inside depends on the range of temperatures over which they need to operate.

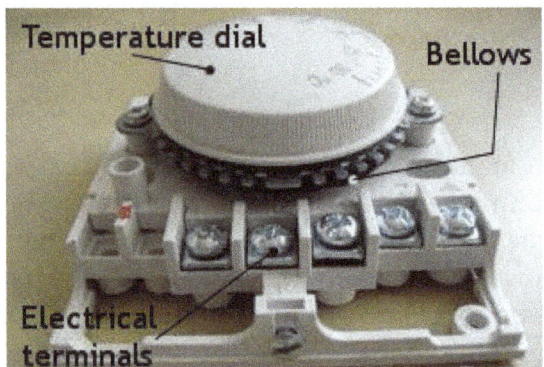 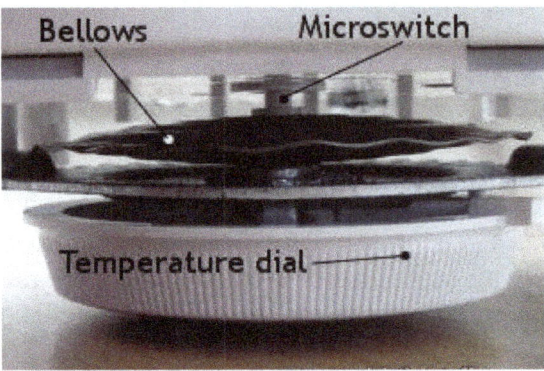

A Honeywell thermostat (the one in our top photo, and shown here with with the case removed, from two different angles) regulates temperature with a pair of metal discs, separated by a gas-filled bellows, that push against a microswitch. Turning the temperature dial moves the discs nearer to or further away from the microswitch in the center. That means the gas bellows has to expand more or less to turn the electricity on or off—effectively raising the temperature at which the switch triggers (and the room temperature).

Wax Thermostats

Summing up what we've discovered already, you can see that all mechanical thermostats (all non-electronic ones) use substances that change size or shape with increasing temperature. So bitmetallic thermostats rely on the expansion of metals as they get hotter, while gas bellows work using the expansion of gases. Some thermostats go further and use the change in state of a substance from liquid to gas. Wax thermostats are probably the most common example—and you'll find them in home radiator valves, car engines and mixer showers. They use a little plug of wax inside a sealed chamber. As the temperature changes, the wax melts (changes state from solid to liquid), expands greatly, and pushes a rod out of the chamber that switches something on or off (operating the engine cooling system in a car or regulating the mixture of hot and cold water in a shower to ensure your body doesn't get boiled like a lobster). Wax thermostats tend to be more reliable and longer lasting in the extreme conditions inside a vehicle engine.

Pneumatic Thermostats

A pneumatic thermostat is a thermostat that controls a heating or cooling system via a series of air-filled control tubes. This "control air" system responds to the pressure changes (due to temperature) in the control tube to activate heating or cooling when required. The control air typically is maintained on "mains" at 15-18 psi (although usually operable up to 20 psi). Pneumatic thermostats typically provide output/branch/post-restrictor (for single-pipe operation) pressures of 3-15 psi which is piped to the end device (valve/damper actuator/pneumatic-electric switch, etc.).

The pneumatic thermostat was invented by Warren Johnson in 1895 soon after he invented the electric thermostat. In 2009, Harry Sim was awarded a patent for a pneumatic-to-digital interface that allows pneumatically controlled buildings to be integrated with building automation systems to provide similar benefits as direct digital control (DDC).

A wax pellet driven valve can be analyzed by graphing the wax pellet's hysteresis which consists of two thermal expansion curves; extension (motion) vs. temperature increase, and contraction (motion) vs. temperature decrease. The spread between the up and down curves visually illustrate the valve's hysteresis; there is always hysteresis within wax driven technology due to the phase change between solids and liquids. Hysteresis can be controlled with specialized blended mixes of hydrocarbons; tight hysteresis is what most desire, however specialized engineering applications require broader ranges. Wax pellet driven valves are used in anti scald, freeze protection, over-temp purge, solar thermal, automotive, and aerospace applications among many others.

Electrical and Analog Electronic Thermostats

Bimetallic Switching Thermostats

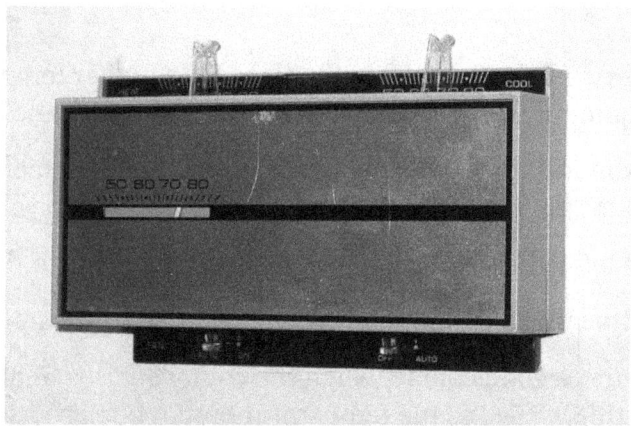

Bimetallic thermostat for buildings

Water and steam based central heating systems have traditionally had overall control by wall-mounted bi-metallic strip thermostats. These sense the air temperature using the differential expansion of two metals to actuate an on/off switch. Typically the central system would be switched on when the temperature drops below the setpoint on the thermostat, and switched off when it rises above, with a few degrees of hysteresis to prevent excessive switching. Bi-metallic sensing is now being superseded by electronic sensors. A principal use of the bi-metallic thermostat today is in individual electric convection heaters, where control is on/off, based on the local air temperature and the setpoint desired by the user. These are also used on air-conditioners, where local control is required.

Simple Two Wire Thermostats

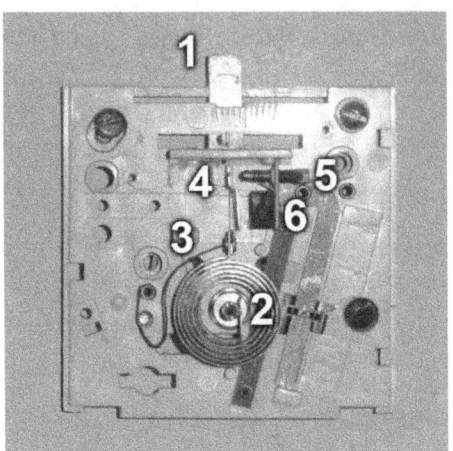

Millivolt thermostat mechanism

The illustration is the interior of a common two wire heat-only household thermostat, used to regulate a gas-fired heater via an electric gas valve. Similar mechanisms may also be used to control oil furnaces, boilers, boiler zone valves, electric attic fans, electric furnaces, electric baseboard heaters, and household appliances such as refrigerators, coffee pots and hair dryers. The power through the thermostat is provided by the heating device and may range from millivolts to 240 volts in common North American construction, and is used to control the heating system either directly (electric baseboard heaters and some electric furnaces) or indirectly (all gas, oil and forced hot water systems). *Due to the variety of possible voltages and currents available at the thermostat, caution must be taken when selecting a replacement device.*

1. Setpoint control lever. This is moved to the right for a higher temperature. The round indicator pin in the center of the second slot shows through a numbered slot in the outer case.

2. Bimetallic strip wound into a coil. The center of the coil is attached to a rotating post attached to lever (a). As the coil gets colder the moving end — carrying (b) — moves clockwise.

3. Flexible wire. The left side is connected via one wire of a pair to the heater control valve.

4. Moving contact attached to the bimetal coil. Thence, to the heater's controller.

5. Magnet. This ensures a good contact when the contact closes. It also provides hysteresis to prevent short heating cycles, as the temperature must be raised several degrees before the contacts will open. As an alternative, some thermostats instead use a mercury switch on

the end of the bimetal coil. The weight of the mercury on the end of the coil tends to keep it there, also preventing short heating cycles. However, this type of thermostat is banned in many countries due to its highly and permanently toxic nature if broken. When replacing these thermostats they must be regarded as chemical waste.

6. Fixed contact screw. This is adjusted by the manufacturer. It is connected electrically by a second wire of the pair to the thermocouple and the heater's electrically operated gas valve.

Millivolt is a separate bimetal thermometer on the outer case to show the actual temperature at the thermostat.

Millivolt Thermostats

As illustrated in the use of the thermostat above, all of the power for the control system is provided by a thermopile which is a combination of many stacked thermocouples, heated by the pilot light. The thermopile produces sufficient electrical power to drive a low-power gas valve, which under control of one or more thermostat switches, in turn controls the input of fuel to the burner.

This type of device is generally considered obsolete as pilot lights can waste a surprising amount of gas (in the same way a dripping faucet can waste a large amount of water over an extended period), and are also no longer used on stoves, but are still to be found in many gas water heaters and gas fireplaces. Their poor efficiency is acceptable in water heaters, since most of the energy "wasted" on the pilot still represents a direct heat gain for the water tank. The Millivolt system also makes it unnecessary for a special electrical circuit to be run to the water heater or furnace; these systems are often completely self-sufficient and can run without any external electrical power supply. For tankless "on demand" water heaters, pilot ignition is preferable because it is faster than hot-surface ignition and more reliable than spark ignition.

Some programmable thermostats - those that offer simple "millivolt" or "two-wire" modes - will control these systems.

24 Volt Thermostats

The majority of modern heating/cooling/heat pump thermostats operate on low voltage (typically 24 volts AC) control circuits. The source of the 24 volt AC power is a control transformer installed as part of the heating/cooling equipment. The advantage of the low voltage control system is the ability to operate multiple electromechanical switching devices such as relays, contactors, and sequencers using inherently safe voltage and current levels. Built into the thermostat is a provision for enhanced temperature control using anticipation. A heat anticipator generates a small amount of additional heat to the sensing element while the heating appliance is operating. This opens the heating contacts slightly early to prevent the space temperature from greatly overshooting the thermostat setting. A mechanical heat anticipator is generally adjustable and should be set to the current flowing in the heating control circuit when the system is operating. A cooling anticipator generates a small amount of additional heat to the sensing element while the cooling appliance is not operating. This causes the contacts to energize the cooling equipment slightly early, preventing the space temperature from climbing excessively. Cooling anticipators are generally non-adjustable.

Electromechanical thermostats use resistance elements as anticipators. Most electronic thermostats use either thermistor devices or integrated logic elements for the anticipation function. In some electronic thermostats, the thermistor anticipator may be located outdoors, providing a variable anticipation depending on the outdoor temperature. Thermostat enhancements include outdoor temperature display, programmability, and system fault indication. While such 24 volt thermostats are incapable of operating a furnace when the mains power fails, most such furnaces require mains power for heated air fans (and often also hot-surface or electronic spark ignition) rendering moot the functionality of the thermostat. In other circumstances such as piloted wall and "gravity" (fanless) floor and central heaters the low voltage system described previously may be capable of remaining functional when electrical power is unavailable.

There are no standards for wiring color codes, but convention has settled on the following terminal codes and colors. In all cases, the manufacturer's instructions should be considered definitive.

Terminal Code	Color	Description
R/V	Red	24 volt
Rh/4	Red	24 volt HEAT load
Rc	Red	24 volt COOL load
C	Black/Blue/Brown	24 volt Common (Ground)
W/W1	White	Heat
W2	Varies/White/Black	2nd Stage/Backup Heat
Y/Y1	Yellow	Cool
Y2	Blue/Orange/Purple/Yellow/White	2nd Stage Cool
G	Green	Fan
O	Varies/Orange/Black	Reversing valve Energize to Cool (Heat Pump)
B	Varies/Blue/Black/Brown/Orange	Reversing valve Energize to Heat (Heat Pump) or Common
E	Varies/Blue/Pink/Gray/Tan	Emergency Heat (Heat Pump)
S1/S2	Brown/Black/Blue	Temperature Sensor (Usually outdoors on a Heat Pump System)
T	Varies/Tan/Gray	Outdoor Anticipator Reset
X	Varies	Emergency Heat (Heat Pump) or Common
X2	Varies	2nd stage/emergency heating or indicator lights
L	Varies	Service Light

Line Voltage Thermostats

Line voltage thermostats are most commonly used for electric space heaters such as a baseboard heater or a direct-wired electric furnace. If a line voltage thermostat is used, system power (in the United States, 120 or 240 volts) is directly switched by the thermostat. With switching current often exceeding 40 amperes, using a low voltage thermostat on a line voltage circuit will result at least in the failure of the thermostat and possibly a fire. Line voltage thermostats are sometimes used in other applications, such as the control of fan-coil (fan powered from line voltage blowing through a coil of tubing which is either heated or cooled by a larger system) units in large systems using centralized boilers and chillers, or to control circulation pumps in hydronic heating applications.

Some programmable thermostats are available to control line-voltage systems. Baseboard heaters will especially benefit from a programmable thermostat which is capable of continuous control effectively controlling the heater like a lamp dimmer, and gradually increasing and decreasing heating to ensure an extremely constant room temperature (continuous control rather than relying on the averaging effects of hysteresis). Systems which include a fan (electric furnaces, wall heaters, etc.) must typically use simple on/off controls.

Digital Electronic Thermostats

Residential digital thermostat

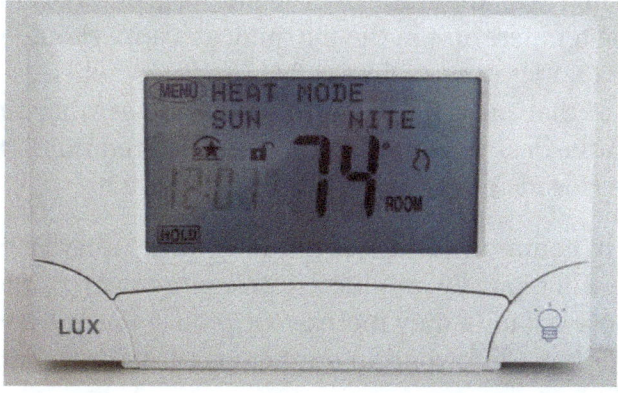

Lux Products' Model TX9000TS Touch Screen Thermostat

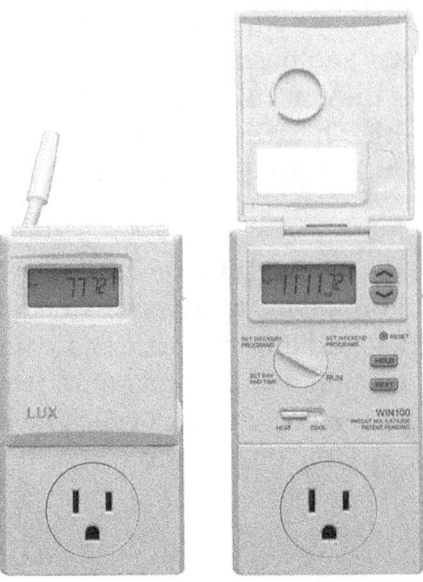

Lux Products WIN100 Heating & Cooling Programmable Outlet
Thermostat shown with control door closed and open

Newer digital thermostats have no moving parts to measure temperature and instead rely on thermistors or other semiconductor devices such as a resistance thermometer (resistance temperature detector). Typically one or more regular batteries must be installed to operate it, although some so-called "power stealing" digital thermostats use the common 24 volt AC circuits as a power source, but will not operate on thermopile powered "millivolt" circuits used in some furnaces. Each has an LCD screen showing the current temperature, and the current setting. Most also have a clock, and time-of-day and even day-of-week settings for the temperature, used for comfort and energy conservation. Some advanced models have touch screens, or the ability to work with home automation or building automation systems.

Digital thermostats use either a relay or a semiconductor device such as triac to act as a switch to control the HVAC unit. Units with relays will operate millivolt systems, but often make an audible "click" noise when switching on or off.

HVAC systems with the ability to modulate their output can be combined with thermostats that have a built-in PID controller to achieve smoother operation. There are also modern thermostats featuring adaptive algorithms to further improve the inertia prone system behaviour. For instance, setting those up so that the temperature in the morning at 7 a.m. should be 21° C (69.8° F), makes sure that at that time the temperature will be 21° C (69.8° F), where a conventional thermostat would just start working at that time. The algorithms decide at what time the system should be activated in order to reach the desired temperature at the desired time. It also makes sure that the temperature is very stable (for instance, by reducing overshoots).

Most digital thermostats in common residential use in North America and Europe are programmable thermostats, which will typically provide a 30% energy savings if left with their default programs; adjustments to these defaults may increase or reduce energy savings. The programmable thermostat article provides basic information on the operation, selection and installation of such a thermostat.

Thermostats and HVAC Operation

Ignition Sequences in Modern Conventional Systems

- Gas

    1. Start drafting fan (if the furnace is relatively recent) to create a column of air flowing up the chimney.

    2. Heat ignitor or start spark-ignition system.

    3. Open gas valve to ignite main burners.

    4. Wait (if furnace is relatively recent) until the heat exchanger is at proper operating temperature before starting main blower fan or circulator pump.

- Oil

    1. Similar to gas, except rather than opening a valve, the furnace will start an oil pump to inject oil into the burner.

- Electric

    1. The blower fan or circulator pump will be started, and a large electromechanical relay or TRIAC will turn on the heating elements.

- Coal (including grains such as corn, wheat, and barley, or pellets made of wood, bark, or cardboard).

    1. Generally rare today (though grains and pellets are increasing in popularity); similar to gas, except rather than opening a valve, the furnace will start a screw to drive coal/grain/pellets into the firebox.

With non-zoned (typical residential, one thermostat for the whole house) systems, when the thermostat's R (or Rh) and W terminals are connected, the furnace will go through its start-up procedure and produce heat.

With zoned systems (some residential, many commercial systems — several thermostats controlling different "zones" in the building), the thermostat will cause small electric motors to open valves or dampers and start the furnace or boiler if it's not already running.

Most programmable thermostats will control these systems.

Combination Heating/cooling Regulation

Depending on what is being controlled, a forced-air air conditioning thermostat generally has an external switch for heat/off/cool, and another on/auto to turn the blower fan on constantly or only when heating and cooling are running. Four wires come to the centrally-located thermostat from the main heating/cooling unit (usually located in a closet, basement, or occasionally in the attic): One wire, usually red, supplies 24 volts AC power to the thermostat, while the other three supply control signals from the thermostat, usually white for heat, yellow for cooling, and green to turn on the blower fan. The power is supplied by a transformer, and when the thermostat makes contact between the 24 volt power and one or two of the other wires, a relay back at the heating/cooling unit activates the corresponding heat/fan/cool function of the unit(s).

A thermostat, when set to "cool", will only turn on when the ambient temperature of the surrounding room is above the set temperature. Thus, if the controlled space has a temperature normally above the desired setting when the heating/cooling system is off, it would be wise to keep the thermostat set to "cool", despite what the temperature is outside. On the other hand, if the temperature of the controlled area falls below the desired degree, then it is advisable to turn the thermostat to "heat".

Heat Pump Regulation

Thermostat design

The heat pump is a refrigeration based appliance which reverses refrigerant flow between the indoor and outdoor coils. This is done by energizing a reversing valve (also known as a "4-way" or "change-over" valve). During cooling, the indoor coil is an evaporator removing heat from the indoor air and transferring it to the outdoor coil where it is rejected to the outdoor air. During heating, the outdoor coil becomes the evaporator and heat is removed from the outdoor air and transferred to the indoor air through the indoor coil. The reversing valve, controlled by the thermostat, causes the change-over from heat to cool. Residential heat pump thermostats generally have an "O" terminal to energize the reversing valve in cooling. Some residential and many commercial heat pump thermostats use a "B" terminal to energize the reversing valve in heating. The heating capacity of a heat pump decreases as outdoor temperatures fall. At some outdoor temperature (called the balance point) the ability of the refrigeration system to transfer heat into the building falls below the heating needs of the building. A typical heat pump is fitted with electric heating elements to supplement the refrigeration heat when the outdoor temperature is below this balance point. Operation of the supplemental heat is controlled by a second stage heating contact in the heat pump thermostat. During heating, the outdoor coil is operating at a temperature below the outdoor temperature and condensation on the coil may take place. This condensation may then freeze onto the coil, reducing its heat transfer capacity. Heat pumps therefore have a provision for occasional defrost of the outdoor coil. This is done by reversing the cycle to the cooling mode, shutting off the outdoor fan, and energizing the electric heating elements. The electric heat in defrost mode is needed to keep the system from blowing cold air inside the building. The elements are then used in the "reheat" function. Although the thermostat may indicate the system is in defrost and electric heat is activated, the defrost function is not controlled by the thermostat. Since the heat pump has electric heat elements for supplemental and reheats, the heat pump thermostat provides for use of the electric heat elements should the refrigeration system fail. This function is normally activated by an "E" terminal on the thermostat. When in emergency heat, the thermostat makes no attempt to operate the compressor or outdoor fan.

Thermostat Location

The thermostat should not be located on an outside wall or where it could be exposed to direct sunlight at any time during the day. It should be located away from the room's cooling or heating vents or device, yet exposed to general airflow from the room(s) to be regulated. An open hallway may be most appropriate for a single zone system, where living rooms and bedrooms are operated as a single zone. If the hallway may be closed by doors from the regulated spaces then these should be left open when the system is in use. If the thermostat is too close to the source controlled then the system will tend to "short a cycle", and numerous starts and stops can be annoying and in some cases shorten equipment life. A multiple zoned system can save considerable energy by regulating individual spaces, allowing unused rooms to vary in temperature by turning off the heating and cooling.

Dummy Thermostats

It has been reported that many thermostats in office buildings are non-functional dummy devices, installed to give tenants' employees an illusion of control. These dummy thermostats are in effect a type of placebo button. However, these thermostats are often used to detect the temperature in the zone, even though their controls are disabled. This function is often referred to as "lockout".

Smart Thermostat

A smart thermostat is also known as a connected or communicating thermostat. It allows you to create automatic and programmable temperature settings based on daily schedules, weather conditions, and heating and cooling needs. Some Wi-Fi thermostats have this advanced functionality, but not all do.

Working of Smart Thermostat

Smart thermostats are comprised of 3 basic components: One part plugs directly into your HVAC system. This part communicates directly with the second component, the thermostat control itself. The third and last part that makes up a 'smart' thermostat is the smart thermostat app, which is downloaded to your smartphone, mobile device or computer. This grand trifecta of tech allows you to adjust home temperatures from any location, from the couch to the coastline, given a wireless or internet connection. While these 'basic' components are what makes a smart thermostat tick, some higher-end models may offer more, such as smart 'learning' models, which require no programming; smartphone tracking, adjusting temperatures as family members get closer to home; or additional sensors, for zoned temperature control and improved energy savings.

Difference of Smart Thermostat from a Standard or Programmable Models

Old Reliable: Analog Thermostats

The cheapest (and oldest) thermostat option, these devices only operate in-home, with the push of your finger. Temperature adjustments and readings are far from exact, hitting you where it hurts: Your energy bill.

A Step-up: Digital Thermostats

These thermostats offer slightly more control and improved temperature readings, with a digital thermometer that turns off your HVAC system when your desired temperature is reached, but providing no programming options.

Good in Theory: Programmable Thermostats

Also digital, programmable thermostat are designed to offer better thermostat control whether you're home or away, allowing for programming based on your schedule. Multiple options exist, including those offering different programming for each day of the week (7-day models), 5-day work week, 2-day weekend scheduling (5+2-day models), and 5+1+1-day models for a varying weekend schedule. Though programmable thermostats are purported to offer energy savings of 10-20%, studies have revealed that is only when homeowners select the right model for their lifestyle, and program it properly, a common issue among programmable thermostat owners.

User-Friendly: Smart Thermostats

Like programmable models, smart thermostats can save you up to 20% on heating and cooling costs, and are an ideal option for those who are unlikely to ever program a thermostat, or do so correctly. Unlike programmable models, they offer remote operation from anywhere via smartphone, mobile device, or voice-operated home automation systems like Google Home and Alexa, and tons of added bells and whistles, including energy-usage reports to help you identify additional opportunities for savings.

Thermostatic Radiator Valve

Thermostatic radiator valves are commonly referred to as TRVs and are used to control the air temperature of different rooms – you will normally find them on the side of your radiators.

TRVs are just one of a number of heating controls, which allow homeowners to heat their homes more efficiently. If set up correctly, they allow you to have different heating zones throughout the house, despite only one centralised boiler providing the heat.

Working Principles of TRV

A Thermostatic Radiator Valve is self-regulating and works by automatically changing the flow of hot water that comes into your radiator at any one time.

It is a pretty simple mechanism to understand and consists of two main parts – the valve headand the valve body – with the head (as you might expect) sitting on top of the body.

When the temperature of the room begins to change, a capsule in the head of the TRV will either expand or contract – automatically moving a pin in the body of the valve which causes the valve to either open or close.

If the temperature in your room drops a little too low, the capsule will contract and pull out the pin; allowing more hot water to enter your radiator and increase its temperature.

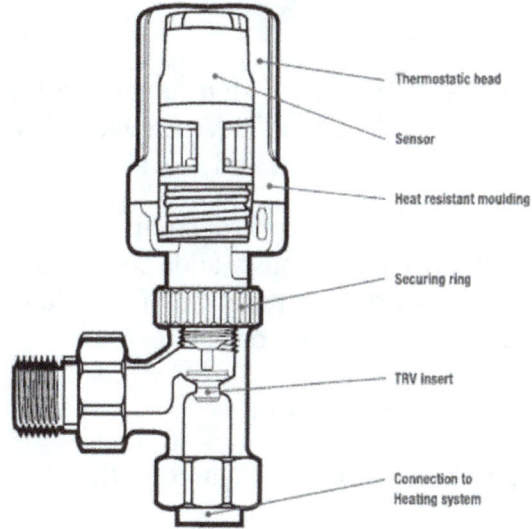

Conversely – if the room begins to get a little too warm – the expansion of the same capsule will cause the pin to close the valve and reduce the amount of hot water.

The capsules in thermostatic radiator valves operate using a metal spring that is filled with wax or liquid – with the liquid type considered to be the best and most consistent at adjusting the temperature.

Place of Installation

There are two places that you really shouldn't install TRVs on the radiators – the first is in bathrooms. This is because the heat produced by the bath/shower will cause the TRV to shut off (it will cause the capsule to expand), just when you need the heat from the radiator to fight off condensation.

We also don't advise installing a TRV in the same room as your main heating thermostat. The main thermostat will link directly to the boiler, firing it up or turning it down, so by having a TRV in that room they will fight for control – if the TRV wins, the heating in your house will go off.

Newer Electronic TRVs

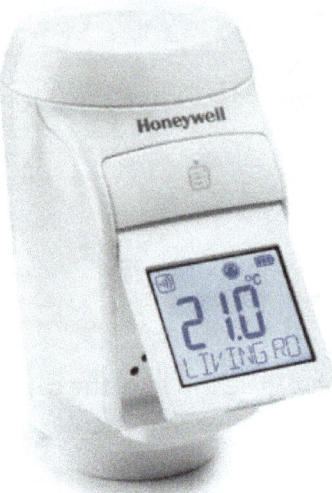

Nowadays you can take zonal heating control a step further with electronic TRVs (for example Honeywell's Evohome system). These electronic TRVs use batteries and electronic thermostats to constantly monitor the temperature of the room and move the pin up and down accordingly.

They can also be used with the other heating control components to create an intelligent heating system like the Heat Genius system. This obviously adds significant cost to the system, but allows you to accurately monitor and control the temperature of individual rooms in the home all from the touch of the button on your tablet computer or phone.

Thermostatic Radiator Valve Problems

TRVs are relatively primitive, especially when compared to the newer smart heating controls that are available on the market.

Their simplicity, though, does mean that it shouldn't be too difficult to diagnose any problems, should they start to fail you, which they sometimes have a tendency to do.

The most common and regularly occurring problem with a thermostatic radiator valve is that the valve – or the head of the valve – begins to stick or catch.

This will leave them open or closed and can happen when the setting on the valve has not been adjusted for some time – such as at the end of a long hot summer.

This is a pretty simple fix and shouldn't result in you having to part with any extra cash – provided that the pin in the valve hasn't failed and the valve head isn't totally immovable.

What you need to do is to turn the TRV to its highest setting – this is usually displayed as a number 5 – to fully open the valve and then remove the top of the TRV by undoing the large thumbwheel just above the radiator tail (this should be pretty easy to do and is most likely not to require a wrench to achieve).

Once removed, you should be able to see a little metal pin that would normally move up an down.

If this happens to be a little stiff – and is preventing the valve from opening an closing correctly – simply give it a spray with some trusty WD40 and work it up and down a few times to try and loosen it up a bit.

If all appears to be ok with the pin, it could be that the wax or liquid cartridge at the top of the TRV has failed and you will need to purchase a new thermostatic radiator valve head.

References

- Ellis, Andrew M.; Mayhew, Christopher A. (2014). Proton Transfer Reaction Mass Spectrometry - Principles and Applications. Chichester, West Sussex, UK: John Wiley & Sons Ltd. ISBN 978-1-405-17668-2.

- How-an-air-flow-meter-works-in-a-vehicle: doityourself.com, Retrieved 20 June 2018

- MartíNez-Hurtado, J. L.; Davidson, C. A. B.; Blyth, J.; Lowe, C. R. (2010). "Holographic Detection of Hydrocarbon Gases and Other Volatile Organic Compounds". Langmuir. 26(19): 15694–9. doi:10.1021/la102693m. PMID 20836549.

- Technically-speaking-what-blower-door-test-tells-you: bpihomeowner.org, Retrieved 23 May 2018

- Matthew Peach, Optics.org. "Photonics-based MINIGAS project yields better gas detectors." Jan 29, 2013. Retrieved Feb 15, 2013.

- How-Gas-Detectors-Work, instruments-controls: thomasnet.com, Retrieved 20 March 2018

- Sandberg, Jared (January 15, 2003). "Employees Only Think They Control Thermostat". The Wall Street Journal. Retrieved September 2, 2009.

- Is-smart-thermostat-worth-buying: efficiencyvermont.com, Retrieved 28 April 2018

- James E. Brumbaugh, AudelHVAC Fundamentals: Volume 2: Heating System Components, Gas and Oil Burners, and Automatic Controls, John Wiley & Sons, 2004 ISBN 0764542079 pp. 109-119

- How-does-a-smart-thermostat-work: mrelectric.com, Retrieved 18 May 2018

- Biasioli, Franco; Yeretzian, Chahan; Märk, Tilmann D.; Dewulf, Jeroen; Van Langenhove, Herman (2011). "Direct-injection mass spectrometry adds the time dimension to (B)VOC analysis". Trends in Analytical Chemistry. 30 (7): 1003–1017. doi:10.1016/j.trac.2011.04.005.

- What-are-thermostatic-radiator-valves: bestheating.com, Retrieved 31 March 2018

- "An Early History Of Comfort Heating". The NEWS Magazine. Troy, Michigan: BNP Media. November 6, 2001. Retrieved November 2, 2014.

6

HVAC Control Systems

In order to control or regulate the heating or air conditioning, we require HVAC control systems. This chapter includes fundamental topics related to building automation, direct digital control systems, electrical and electronic control systems, microprocessor systems, etc. for a complete understanding of HVAC control systems.

The capacity of the HVAC system is typically designed for the extreme conditions. Most operation is part load/off design as variables such as solar loads, occupancy, ambient temperatures, equipment & lighting loads etc keep on changing through out the day. Deviation from design shall result in drastic swings or imbalance since design capacity is greater than the actual load in most operating scenarios. Without control system, the system will become unstable and HVAC would overheat or overcool spaces.

Control Strategies

The simplest control in HVAC system is cycling or on/off control to meet part load conditions. If building only needs half the energy that the system is designed to deliver, the system runs for about ten minutes, turns off for ten minutes, and then cycles on again. As the building load increases, the system runs longer and its off period is shorter.

One problem faced by this type of control is short-cycling which keeps the system operating at the inefficient condition and wears the component quickly. A furnace or air-conditioner takes several minutes before reaching "steady-state" performance.

The longer the time between cycles, the wider the temperature swings in the space. Trying to find a compromise that allows adequate comfort without excessive wear on the equipment is modulation or proportional control. Under this concept, if a building is calling for half the rated capacity of the chiller, the chilled water is supplied at half the rate or in case of heating furnace; fuel is fed to the furnace at half the design rate: the energy delivery is proportional to the energy demand. While this system is better than cycling, it also has its problems. Equipment has a limited turn-down ratio. A furnace with a 5:1 turn-down ratio can only be operated above 20% of rated capacity. If the building demand is lower than that, cycling would still have to be used.

An alternate method of control under part-load conditions is staging. Several small units (e.g., four units at 25% each) are installed instead of one large unit. When conditions call for half the design capacity, only two units operate. At 60% load, two units are base-loaded (run continuously), and a third unit swings (is either cycled or modulated) as needed. To prevent excessive wear, sequencing is often used to periodically change the unit being cycled.

The HVAC control system is typically distributed across three areas:

1. The HVAC equipment and their controls located in the main mechanical room. Equipment includes chillers, boiler, hot water generator, heat exchangers, pumps etc.

2. The weather maker or the "Air Handling Units (AHUs)" may heat, cool, humidify, dehumidify, ventilate, or filter the air and then distribute that air to a section of the building. AHUs are available in various configurations and can be placed in a dedicated room called secondary equipment room or may be located in an open area such as roof top air-handling units.

3. The individual room controls depending on the HVAC system design. The equipment includes fan coil units, variable air volume systems, terminal reheat, unit ventilators, exhausters, zone temperature/humidistat devices etc.

Benefits of a Control System

Controls are required for one or more of the following reasons:

1. Maintain thermal comfort conditions.

2. Maintain optimum indoor air quality.

3. Reduce energy use.

4. Safe plant operation.

5. To reduce manpower costs.

6. Identify maintenance problems.

7. Efficient plant operation to match the load.

8. Monitoring system performance.

Building Automation

Complete autonomous control of an entire facility is the goal that any modern automation system attempts to achieve. The distributed control system - the computer networking of electronic devices designed to monitor and control the mechanical, security, fire, lighting, HVAC and humidity control and ventilation systems in a building or across several campuses.

The Building Automation System (BAS) core functionality is to keep building climate within a specified range, light rooms based on an occupancy schedule, monitor performance and device failures in all systems and provide malfunction alarms. Automation systems reduce building energy and maintenance costs compared to a non-controlled building. Typically they are financed through energy and insurance savings and other savings associated with pre-emptive maintenance and quick detection of issues.

A building controlled by a BAS is often referred to as an intelligent building or "smart building". Commercial and industrial buildings have historically relied on robust proven protocols like BACnet.

Almost all multi-story green buildings are designed to accommodate a BAS for the energy, air and water conservation characteristics. Electrical device demand response is a typical function of a BAS, as is the more sophisticated ventilation and humidity monitoring required of "tight" insulated buildings. Most green buildings also use as many low-power DC devices as possible, typically integrated with power over Ethernet wiring, so by definition always accessible to a BAS through the Ethernet connectivity. Even a passivhaus design intended to consume no net energy whatsoever will typically require a BAS to manage heat capture, shading and venting, and scheduling device use.

A Simple Example of an Intergrated Building Automation System

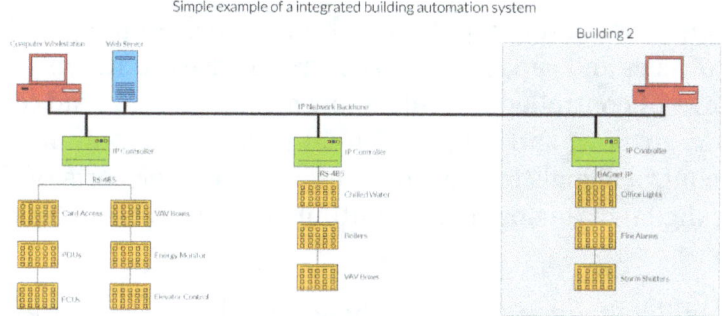

Simple example of a integrated building automation system

Communication of a BAS

Buses and Protocols

Most building automation networks consist of a primary and secondary bus which connect high-level controllers with lower-level controllers, input/output devices and a user interface devices. ASHRAE's open protocol BACnet or the open protocol LonTalk specify how most such devices interoperate. Modern systems use SNMP to track events, building on decades of history with SNMP-based protocols in the computer networking world.

Physical connectivity between devices was historically provided by dedicated optical fiber, ethernet, ARCNET, RS-232, RS-485 or a low-bandwidth special purpose wireless network. Modern systems rely on standards-based multi-protocol heterogeneous networking. These accommodate typically only IP-based networking but can make use of any existing wiring, and also integrate powerline networking over AC circuits, power over Ethernet low power DC circuits, high-bandwidth wireless networks such as LTE and IEEE 802.11n and IEEE 802.11ac and often integrate these using the building-specific wireless mesh open standards.

Current systems provide interoperability at the application level, allowing users to mix-and-match devices from different manufacturers, and to provide integration with other compatible building control systems. These typically rely on SNMP, long used for this same purpose to integrate diverse computer networking devices into one coherent network.

Types of Inputs and Outputs

Analog inputs are used to read a variable measurement. Examples are temperature, humidity and pressure sensors. A digital input indicates if a device is turned on or not. Analog outputs control the speed or position of a device, such as a variable frequency drive or a valve or damper actuator. Digital outputs are used to open and close relays and switches. An example would be to turn on the parking lot lights when a photocell indicates it is dark outside.

Devices

Controllers

BAS Controllers are purpose-built computers with input and output capabilities. These controllers come in a range of sizes and capabilities to control devices commonly found in buildings and to control sub-networks of controllers. Inputs allow a controller to read temperatures, humidity, pressure, current flow, air flow, and other essential factors. The outputs allow the controller to send command and control signals to slave devices, and to other parts of the system. Inputs and outputs can be either digital or analog. Digital outputs are also sometimes called discrete depending on manufacturer.

Controllers used for building automation can be grouped in 3 categories. Programmable Logic Controllers (PLCs), System/Network controllers, and Terminal Unit controllers. However an additional device can also exist in order to integrate 3rd party systems (i.e. a stand-alone AC system) into a central Building automation system).

System/Network controllers may be applied to control one or more mechanical systems such as an Air Handler Unit (AHU), boiler, chiller, etc., or they may supervise a sub-network of controllers. In the diagram above, System/Network controllers are often used on the IP backbone.

Terminal Unit controllers usually are suited for control of lighting and simpler devices such as a package rooftop unit, heat pump, VAV box, or fan coil, etc. The installer typically selects 1 of the available pre-programmed personalities best suited to the device to be controlled, and does not have to create new control logic.

Air Handlers

Most air handlers mix return and outside air so less temperature/humidity conditioning is needed. This can save money by using less chilled or heated water (not all AHUs use chilled/hot water circuits). Some external air is needed to keep the building's air healthy. To optimize energy efficiency while maintaining healthy indoor air quality (IAQ), demand control (or controlled) ventilation (DCV) adjusts the amount of outside air based on measured levels of occupancy. Analog or digital temperature sensors may be placed in the space or room, the return and supply air ducts, and sometimes the external air. Actuators are placed on the hot and chilled water valves, the outside air and return air dampers. The supply fan (and return if applicable) is started and stopped based on either time of day, temperatures, building pressures or a combination.

Constant Volume Air-handling Units

The less efficient type of air-handler is a "constant volume air handling unit," or CAV. The fans in CAVs do not have variable-speed controls. Instead, CAVs open and close dampers and water-supply valves to maintain temperatures in the building's spaces. They heat or cool the spaces by opening or closing chilled or hot water valves that feed their internal heat exchangers. Generally one CAV serves several spaces

Variable Volume Air-handling Units

A more efficient unit is a "variable air volume (VAV) air-handling unit," or VAV. VAVs supply pressurized air to VAV boxes, usually one box per room or area. A VAV air handler can change the pressure to the VAV boxes by changing the speed of a fan or blower with a variable frequency drive. The amount of air is determined by the needs of the spaces served by the VAV boxes.

Each VAV box supply air to a small space, like an office. Each box has a damper that is opened or closed based on how much heating or cooling is required in its space. The more boxes are open, the more air is required, and a greater amount of air is supplied by the VAV air-handling unit. Some VAV boxes also have hot water valves and an internal heat exchanger. The valves for hot and cold water are opened or closed based on the heat demand for the spaces it is supplying. These heated VAV boxes are sometimes used on the perimeter only and the interior zones are cooling only. A minimum and maximum CFM must be set on VAV boxes to assure adequate ventilation and proper air balance.

Chilled Water System

Chilled water is often used to cool a building's air and equipment. The chilled water system will have chiller(s) and pumps. Analog temperature sensors measure the chilled water supply and return lines. The chillers are sequenced on and off to chill the chilled water supply.

A chiller is a refrigeration unit designed to produce cool (chilled) water for space cooling purposes. The chilled water is then circulated to one or more cooling coils located in air handling units, fan-coils, or induction units. Chilled water distribution is not constrained by the 100 foot separation limit that applies to DX systems, thus chilled water-based cooling systems are typically used in larger buildings. Capacity control in a chilled water system is usually achieved through modulation of water flow through the coils; thus, multiple coils may be served from a single chiller without compromising control of any individual unit. Chillers may operate on either the vapor compression principle or the absorption principle. Vapor compression chillers may utilize reciprocating, centrifugal, screw, or rotary compressor configurations. Reciprocating chillers are commonly used for capacities below 200 tons; centrifugal chillers are normally used to provide higher capacities; rotary and screw chillers are less commonly used, but are not rare. Heat rejection from a chiller may be by way of an air-cooled condenser or a cooling tower (both discussed below). Vapor compression chillers may be bundled with an air-cooled condenser to provide a packaged chiller, which would be installed outside of the building envelope. Vapor compression chillers may also be designed to be installed separate from the condensing unit; normally such a chiller would be installed in an enclosed central plant space. Absorption chillers are designed to be installed separate from the condensing unit.

Hot Water System

The hot water system supplies heat to the building's air-handling unit or VAV box heating coils, along with the domestic hot water heating coils (Calorifier). The hot water system will have a boiler(s) and pumps. Analog temperature sensors are placed in the hot water supply and return lines. Some type of mixing valve is usually used to control the heating water loop temperature. The boiler(s) and pumps are sequenced on and off to maintain supply.

Direct Digital Control Systems

Direct Digital Controls are a computerized program that monitors and controls all of the HVAC units within a commercial building to provide a complete energy management system. DDC makes it possible to view each unit on a computer monitor and see how each unit is performing ay any given time. If the unit experiences a problem, the system has the capability to page a service technician. DDC systems can also control lighting, alarms and fire detection equipment.

Direct Digital Controls have become a necessary component for HVAC systems seeking to achieve maximum energy efficiency. Our experience with these advanced systems allows us to provide effective solutions for commercial building of all sizes.

Structure of Large DDC System

A smaller building may have a single, computerized HVAC controller that operates the direct digital control (DDC) system. However, a large building normally requires a more complex system of controllers, divided into separate sections called tiers. This is called the architecture of the system.

First Tier

The first tier is a central workstation, consisting of a dedicated computer, monitor, keyboard, and printer. Sometimes the keyboard is omitted at the central workstation and a laptop computer is plugged in when instructions are given to the dedicated computer. The central workstation is also called an operator-machine interface (OMI) or user interface (UI). It communicates with the controllers on the second and third tiers. The central workstation can receive, process, store, send, and print data.

The technician can communicate with the central computer by using a keyboard, mouse, touch screen, monitor, and printer. By using these, the technician can:

Workstation for a DDC system. (Courtesy ASI Controls.)

- Check the status of each component and the status of the system.

- See trends that indicate potential system problems, such as a gradually increasing difference between a space temperature and the set point.

- Change settings:

  - Some settings, such as set points and times of operation, can be changed by the building operator or HVAC technician.

  - Other settings, those that would have major effects on the system operation, require a special password to access and change.

Second Tier

The second tier is a system controller.

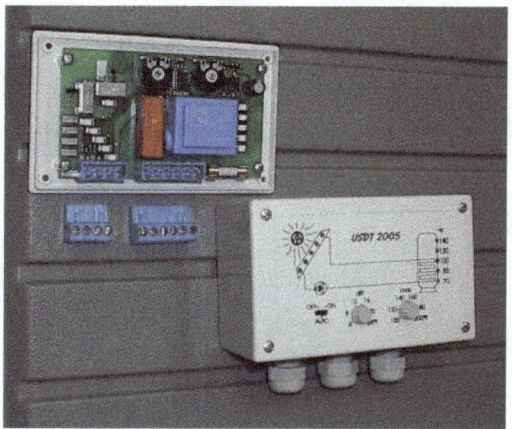

Terminal controller for a boiler

Third Tier

The third tier is a series of terminal controllers. Each control panel is programmed to control one major piece of equipment such as a VAV terminal, chiller, cooling tower, or boiler.

Tiers Communicate

The controllers on different tiers communicate with each other:

- The terminal controllers communicate with the second tier controller.

- The second tier controller communicates with the central workstation.

- The central workstation can display or print out operating data for any component of a system.

This means that you can get information about any component from the central computer in order to identify problems and their probable cause. This takes much less time than it would take to test individual components in the mechanical equipment room.

An HVAC controller in a mechanical room

Working of a Controller

In a DDC system, a controller is a dedicated computer, which means it is designed to operate only one specific program. An HVAC controller receives information from sensors and sends signals to actuators. It is often located in the mechanical equipment room.

Figure below shows a typical single zone DDC system. The controller in figure below has a program to control the air handler portion of an HVAC system. It receives digital or analog input data, processes it, and sends appropriate digital or analog output signals to the HVAC system.

In a DDC system:

- Temperature sensors are electronic. They are either:

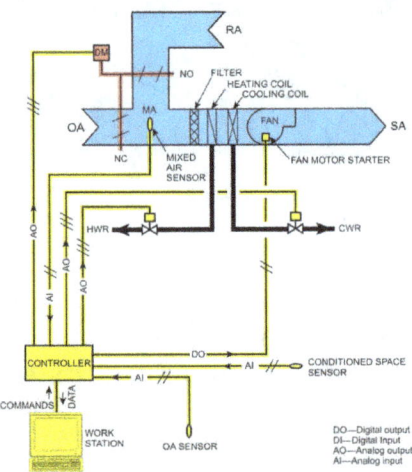

Input and output signals for an HVAC controller for a single zone system

  - RTDs (resistance temperature devices): Electric resistance increases as temperature increases.

  - Thermistors: Electric resistance decreases as temperature increases.

- Valve and damper actuators are electrically operated.

- Both the input and output signals to and from the controller are electric, instead of pneumatic. They are in volts, milliamps, or ohms. As shown in figure above, the signals can be:

  - AI (analog input), such as signals from temperature sensors to the controller.

- AO (analog output), such as signals from the controller to damper or valve actuators.

- DI (digital input), such as a signal from the fan motor starter to the controller to indicate whether the fan is running.

- DO (digital output), such as a signal from the controller to start and stop a fan motor.

Desktop computer connecting to a DDC system.

Communicating with the System

Every manufacturer of an HVAC digital control system uses the DDC to perform pretty much the same functions. The catch is that each DDC system has different ways of letting you use these functions depending on the computer program being used. If you are used to operating computers, you have the basic knowledge required. Most systems use Windows - which is known by most computer users - as the operating system. But each system uses different commands, keystrokes, mouse clicks, and icons. You have to study the instructions for your particular program.

The technician communicates with controllers in the system by using some sort of operator-machine interface (OMI). Many different devices are used as operator-machine interfaces. The device to be used depends on how complex the system is and what the technician needs. The following are different OMIs.

Desktop Computer

Large systems, such as complete building automation systems, or networks that contain more than one building, usually have a desktop computer at the central workstation. It allows a qualified operator with an access code to make system-wide changes in the DDC system. Technicians and building operators who do not have the access code are limited to receiving data, making changes in times and set points, and reading temperatures and airflow rates.

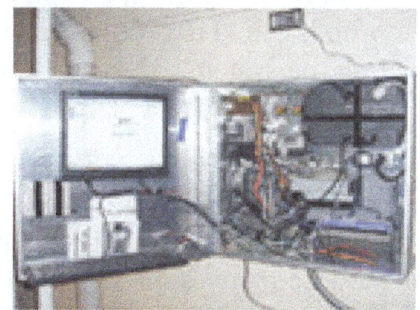

Laptop computer connecter to a controller

Laptop Computer

Building managers and building operators often use a laptop computer plugged into the main terminal. Laptop computers are also mounted in a controller. The laptop can perform the same functions as the desktop computer.

Handheld Terminal

For small systems, a handheld terminal is often used. This is a small computer designed for only this one purpose. It has a limited number of display keys and does not have a keyboard. It displays only two to four lines of data. The handheld terminal provides limited operating data and generally allows for changing only times and set points.

Keypad

A keypad may be permanently mounted on a control panel. Keypads usually perform the same functions as handheld terminals. A keypad has only a few keys and displays two to four lines of data.

Handheld terminal

Dumb Terminals

A dumb terminal is a display and keyboard similar to a keypad or handheld terminal. However, it cannot be used to change settings. Dumb terminals usually display 20 lines or more of data.

Outside Access

Many DDC systems can be accessed from phone lines, cell phone connections, or Internet connections. For example, a technician who is called to take care of a problem in the system can use a cell phone or regular phone to dial into the system. Or the technician can use a desktop or laptop computer to access the computer through the Internet.

Outside access can be used to change such values as set points and time schedules, and to control some alarms. By pressing the proper sequence of numbers on the phone, the system can be accessed and changes can be made. Each system has a specific sequence of numbers to perform an operation. The instructions for these sequences may be by voice on the phone or can be on a printed card.

Keypad

Alarms

Alarm printers are standard computer printers that are operated by the DDC software program. They print out alarms when values in the DDC system exceed a preset amount. Often the phone number of the person to notify is printed with a description of the alarm. The software program may also allow printing preventive maintenance instructions.

Some systems have a pager system. When an alarm occurs on a system, an on-call technician is automatically contacted through a paging service. Some systems use a beeper; others show a short message on a small display. The on-call technician - either on-site or from an outside access - can use an OMI to enter the system and solve the problem.

Some systems use the phone system to communicate. For example, a system may be programmed to signal an alarm by dialing a technician or an office at an off-site location. The phone number called can vary according to the time of day and the person on duty. Typical off-site locations are:

- The office for a building manager for one building or for a number of buildings.

- The office of a service or controls contractor responsible for a number of small commercial establishments (such as restaurants or medical offices).

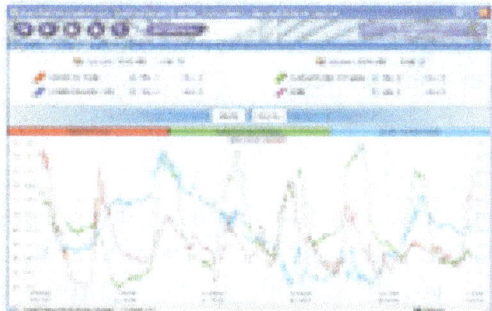

A display showing trends for outside air, relative humidity, and carbon dioxide

Functions

Because a DDC system is computer controlled, it can perform complex calculations and store a vast amount of data. It can monitor trends and make many kinds of adjustments. Some common DDC operations for HVAC are listed below. Depending on the size of the system there can be many more.

Scheduling

In addition to the ordinary schedule for the building's occupied times, DDC systems can control many different time schedules, such as the following:

- Precooling or preheating to bring the building to comfort level by the time of occupancy. The time of starting will vary as the computer calculates different variables such as outside and inside air temperatures to determine the best start time.

- Stop control turns off the HVAC system a set time before the building becomes unoccupied. For an office building, stopping the system a half-hour early means reducing the operating time of the system by 130 hours per year.

- Different set points for unoccupied times.

- Variable time schedules for different days of the week. For example, in addition to the regular daytime occupancy, a building may be in use three nights a week between 7 and 10 p.m.

- Holiday scheduling.

- Daylight savings time adjustment.

- Temporary scheduling - A timed override switch can be used for areas that have variable schedules. Monthly or weekly events can be programmed through an interface.

Economizer Control

Economizer control sets the system to use outside air instead of the mechanical chiller to cool the building, when outside air temperature and humidity are within a specified range.

Sequence Control

Sequence control, also called lead/lag control, is used if two or more pieces of the same item of equipment (such as pumps, chillers, or compressors) are installed in a system. Sequence control puts one in operation if the other fails. For example, if there is a primary pump and a backup pump, if the primary pump fails, the backup is started. Another strategy is to alternate the use of the two pumps so that they receive equal wear.

Reset Control

Reset control is an energy-saving strategy that changes the set point of a controlled variable as another variable changes. For example, a typical hot water set point for a heating coil is 220° F when the outside air is at 0°. When the outside air temperature reaches 60°, the hot water set point could be automatically reduced to 140°, because less energy is needed to maintain the air temperature.

Low Limit and High Limit Control

Low limit and high limit control adjusts the system to limit the low and high of a controlled variable. For example, if the outside air temperature drops very low, the system adjusts to prevent the mixed air temperature from dropping below a certain set point.

Electrical Demand Control

In addition to the regular usage rate, most utility companies impose a demand charge. This is an extra charge per kilowatt based on the highest rate of use for a given period (usually 15 minutes). The demand charge can amount to as much as half the utility bill. DDC programs limit the demand charge in a number of different ways:

- Duty cycling - This turns off different HVAC units during peak load times in order to reduce the demand charge. The system regulates off time according to such things as indoor temperature.

- Load shedding - This turns off various building electrical loads to reduce demand charges. Usually a schedule called a shed table is developed that sets priorities on which loads will be first to be turned off.

- Averaging control - This averages the input from several sensors. For example, building temperatures will vary in a building. The temperature in a lobby will be much different than the temperature of an inner office on one of the higher floors. The computer receives input from sensors located in different parts of the building, averages them, and acts on the average.

- Soft starting - This brings large equipment on line slowly to minimize the large onrush of current that occurs when starting large motors.

DDC and Building Automation

If you are responsible for a building HVAC system, you are primarily concerned with DDC for HVAC. However, the DDC system that controls HVAC may also control many other building systems. The trend in DDC is toward total automation of all building energized systems. In addition to controlling the HVAC system, one central computer often controls scheduling, data gathering, monitoring, and identifying trouble spots for all building functions such as:

- Life safety.
- Fire protection.
- Security.
- Energy management.
- Lighting schedules.
- Equipment monitoring and maintenance.

With a complete building DDC control system, a technician can manage all building functions from one workstation. On-off schedules for lighting and equipment can be set; daily schedules can be altered; and input data can be obtained. The computer may be able to analyze trends and automatically change set points for many systems and conditions.

Benefits of Direct Digital Control System

Convenient Automation of Security Systems

Motion sensors connected to the Direct Digital Controls can turn on lights when a car pulls into the parking garage or parking lot in order to provide better illumination and safety for the occupants of the building. If the system utilizes keyfobs for residents, the resident that just entered onto the property may be able to unlock the exterior door by waving the keyfob in front of a scanner. This eliminates the need to locate keys and manually unlock the door, which increases the speed at which the resident can get into the building.

Cost Effective

With the right sensors and software in place, building and property managers can quickly view the energy efficiency of the building and its components. For example, a maintenance technician or the property manager may notice that the climate control system and lights are on in certain sections of the building when they are least likely to be occupied. As a result, the individual can set the lighting to turn off and on at certain times and program the thermostat in that zone of the building to lower or raise the temperature in order to save money and make the building more energy efficient.

Easier Maintenance

Sensors on HVAC systems, mechanical ventilation and on the electric system can send alerts when a parameter is outside of the normal or expected range. Once the alert has been noticed by management, they can take steps to remedy the problem before it becomes critical or results in the failure of a system.

Easy Climate and Lighting Control

DDC controls make it possible to set all the climate control zones and lighting zones from any computer that contains the DDC control software. This is much faster than locating all of the thermostats and light controls in every section of the building.

Retrocommissioning Made Simple

Retrocommissioning can involve changing out hardware, which is a time-consuming, expensive process. With the right type of DDC Controls, the process simply involves changing parameters in the software or updating the software package so that it can provide more detailed information. Another benefit is that once the software changes have been made, observation of the system can yield vital data on its functionality within a relatively short period of time.

Electrical and Electronic Control Systems

An electrical control system is a physical interconnection of devices that influences the behaviour of other devices or systems. A simple electronic system is made up of an input, a process, and an output. Both input and output variables to the system are signals. Examples of such systems include circulation pumps, compressors, manufacturing systems, refrigerationplant and motor control panels.

Input devices such as sensors gather and respond to information and control a physical process by using electrical energy in the form of an output action. Electronic systems can be classed as 'causal' in nature. The input signal is the 'cause' of the change in the process or system operation, while the output signal is the 'effect', the consequence of the cause. An example is a microphone (input device) causing sound waves to be converted into electrical signals and being amplified by a speaker (output device) producing sound waves.

Electronic systems are commonly represented as a series of interconnected blocks and signals. Each block is shown with its own set of inputs and outputs. This is known as block-diagram representation.

Electrical systems operate either on continuous-time (CT) signals or discrete-time (DT) signals.

A CT system is where the input signals are continuous over time. These tend to be analogue systems producing a linear operation with input and output signals referenced over a set time period, such as between 13:00 and 14:00.

A DT system is where the input signals are a sequence or series of signal values defined in specific time intervals, such as 13:00 and 14:00 separately.

Control systems are one of two different types, either an open loop system or a closed loop system.

Open Loop Control System

An open loop control system is one in which the output does not feedback to the input to correct variations. Instead, the output is varied by varying the input. This means that external conditions will not impact on the system output. An example is a timer-controlled central heating boiler which is switched on between certain preset times regardless of the thermal comfort level of the building.

The benefits of open loop systems are that they are simple, easy to construct and generally remain stable. However, they can be inaccurate and unreliable due to the output not being corrected automatically.

Closed Loop Control System

A closed loop control system is one in which the output has an effect upon the input to maintain a desired output value. It achieves this by providing a feedback loop. For example, a boilermay have a temperature thermostat which monitors the thermal comfort level of a building and sends a feedback signal to ensure the controller maintains the set temperature.

Closed loop systems have the advantage of being accurate, and can be made more or less sensitive depending on the required stability of the system. However, they are more complex in terms of designing a stable system.

Types of Control

There are a number of different types of control:

Manual Control

This system uses no automatic controls, the link is provided by the human operator.

Semi-automatic Control

A sequence of operations is carried out automatically after being started by a human operator. An example is starting an electric motor.

Automatic Control

The human operator is replaced by a controller which monitors the system in comparison with a desired value, using feedback loops to take corrective action if necessary.

Local Control

A level, hand wheel or other attachment fixed on the unit 'locally' is used as a means of alteration and control.

Remote Control

The regulating unit is connected to an actuating device mounted some distance away by means of power transmission through electrical linkages. For example, a remote control to turn on an air-conditioning unit.

On/off Control

The regulating unit can occupy only one of the two available positions of 'on' or 'off'. An example is the on-off switch for lights.

Step-by-step Control

More than two positions can be occupied by the regulating unit but the action occurs in stepsrather than being continuous.

Microprocessor Systems

The sensors and output devices (e.g., actuators, relays) used for electronic control systems are usually the same ones used on microprocessor-based systems. The distinction between electronic control systems and microprocessor-based systems is in the handling of the input signals. In an electronic control system, the analog sensor signal is amplified, and then compared to a set point or override signal through voltage or current comparison and control circuits. In a microprocessor-based system, the sensor input is converted to a digital form, where discrete instructions (algorithms) perform the process of comparison and control. Most subsystems, from VAV boxes to boilers and chillers, now have an onboard DDC system to optimize the performance of that unit. A communication protocol known as BACNet is a standard protocol that allows control units from different manufacturers to pass data to each other.

A patent pending microprocessor controller, detection algorithm and low cost sensor have been developed to accurately detect a refrigerant low charge state in mobile air conditioning systems. The device is intended to prevent damage to the air conditioning system compressor and was designed primarily for truck and bus applications although the principles may also be applied to automotive, agricultural, construction, mining equipment and stationary HVAC applications. The system uses a single, thermistor based refrigerant charge sensor, an optional high pressure transducer and a microprocessor controlled electronics module coupled with standard air conditioning system components to detect the presence or lack of refrigerant in the A/C circuit.

These unique processors have been indegnously developed to control entire referigeration plants with multiple compressors, fans and motors. With muti-point temperature sensing (upto 10 NTC sensor inputs), 8 nos. 4-20 ma loops, and 22 optically isolated feedbacks, the system is a comprehensive solution for controlling industrial Air Conditioning/referigeration requirements. it has extensive datalogging capabilities and can be monitored over Rs. 485 network, Ethernet networks and are capable of being monitored and controlled over GSM networks.

Pneumatic Control Systems

Pneumatics is a branch of engineering that makes use of gas or pressurized air. In engineering, pneumatic control systems can be an effective and economical choice for those designing commercial buildings, yet it isn't used as much as it once was.

Working of Pneumatic Control System

A pneumatic control system uses compressed air as a method of control for HVAC systems. Compressed air is carried via copper and plastic tubes from a controller to a control device, usually a damper or valve actuator. This control method relies on sensors and thermostats that bleed or retain the line pressure from the sensor to the control device and the actuator. Each senor responds to changes in temperature, humidity, and static pressure as examples, to provide feedback in a control loop to open or close the actuator to meet the control set point. The actuators contain diaphragms and spring to function in sequence with the control signal. This system uses the compressed air as the communication method. Each thermostat in a building with a pneumatic control system has one or more air lines connected to it from the main source of compressed air and to some type of final device such as a valve.

Reasons to use Pneumatics

Actuators

Pneumatic actuators, the air-powered "motors" which control valves and dampers, remain the most reliable, durable, and economical actuators available. They require essentially no maintenance except for inspection and adjustment of the mechanical linkages. It is easy to select actuators which fail in the desired position upon loss of electric power (and thus pneumatic air pressure). Pneumatic actuators, of all sizes, cost much less than corresponding electric/electronic actuators.

Modular Control Components

The complete pneumatic control system "brains" are distributed throughout the building using numerous pneumatic building blocks such as thermostats, receiver controllers, and pneumatic relays. Virtually any control strategy can be implemented by a custom-designed pneumatic control system. Pneumatics can provide excellent control performance and can maintain setpoints accurately.

Limitations of Pneumatics

By its own design, the method of using pneumatics is not overly complicated, however there is a tendency of modern building managers to overlook the need for pneumatic maintenance, or to mistakenly believe that the same person who sweeps the floors can maintain the controls. Pneumatics require two important types of maintenance. The first is to ensure that the pneumatic air supply is clean, reliable, and dry at all times. The second is to carry out a program of routine and preventative maintenance under which all control components are inspected and, if necessary, adjusted every couple of years or so. (Left unattended, pneumatic controllers and thermostats may eventually drift away from setpoint). For many commercial buildings, the best way to maintain pneumatic controls systems is to have a service contract with an outside company.

Whilst the technology continues to be used successfully, operational systems are functioning with components that are 40 to 50 years old and sometimes given their age, these systems are deemed to be un-serviceable and there are many cases where pneumatic systems have been pulled out and replaced because of perceived poor performance.

Pneumatic controls are generally replaced by more modern DDC (Direct Digital Controls) based systems partly because they are seen to be a more antiquated and less effective method of control, yet a lot of the issues with pneumatics stems from poor maintenance rather with the ability of the technology.

Sensors for Control

Sensors and analyzers are a control system's window to the world. A sensor is defined as a device that converts a physical stimulus into a readable output, and the definition is illustrated with several examples of engineered and biological sensors. The design of sensors is driven by desired improvements on one or more of surprisingly many performance features and attributes: signal-to-noise ratio, reliability, safety and intrinsic safety, accuracy, response time, dynamic range, cost, power consumption, size, electromagnetic interference immunity, etc.

Control systems include sensors and actuators, the critical pieces needed to ensure that our automation systems can help us manage our activities and environments in desired ways. By extracting information from the physical world, sensors provide inputs to control and automation systems.

The role of a sensor in a simple automation system is depicted in figure. The detection and measurement of some physical effect provides information to the control system regarding a related property of the system under control, which we are interested in regulating to within some 'set point' range.

The controller outputs a command to an actuator (a valve, for example) to correct for measured deviations from the set point, and the control loop is thereby closed.

Because of the simplicity of the control system example of figure, it represents a fair number of practical control systems. In especially simple systems, a distinct controller may not be immediately evident. For example, the 'Honeywell Round' thermostat contains a bimetal strip as an analog sensing mechanism that responds to temperature, and the switch attached to it serves as the actuator.

This integration of sensor and actuator turns a furnace or other space conditioning device on or off, depending on whether the room temperature is within the set-point differential.

In general, however, the trend is to incorporate more, not less, information processing with the sensor. The increasing complexity of sensors is in part a consequence of this trend. In many cases, the information processing is being incorporated within the sensor device, blurring the distinction between transducer and processor, and between sensor and instrument.

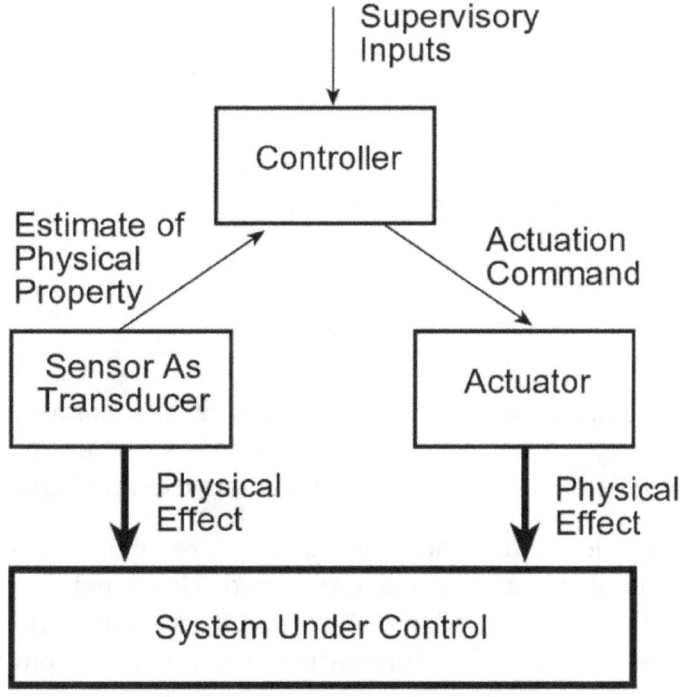

Example of a simple control system

Temperature Sensors

A temperature sensor is a device, usually an RTD (resistance temperature detector) or a thermocouple, that collects the data about temperature from a particular source and converts the data into understandable form for a device or an observer. Temperature sensors are used in many applications like HVand AC system environmental controls, food processing units, medical devices, chemical handling and automotive under the hood monitoring and controlling systems, etc.

The most common type of temperature sensor is a thermometer, which is used to measure temperature of solids, liquids and gases. It is also a common type of temperature sensor mostly used for non-scientific purposes because it is not so accurate.

Types

There are different types of temperature sensors that have sensing capacity depending upon their range of application. Different types of temperature sensors are as follows:

- Thermocouples.
- Resistor temperature detectors.
- Thermistors.
- Infrared sensors.
- Semiconductors.
- Thermometers.

Thermocouples

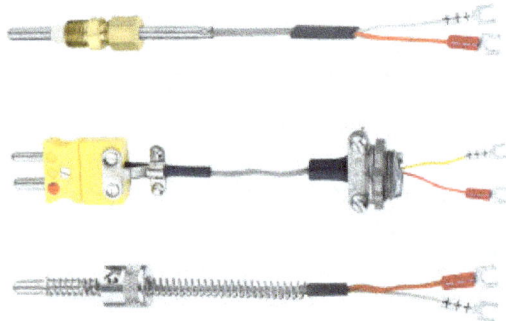

Thermocouple sensor is the most commonly used temperature sensor and it is abbreviated as TC. This sensor is extremely rugged, low-cost, self-powered and can be used for long distance. There are many types of temperature sensors that have a wide range of applications.

A thermocouple is a voltage device that indicates temperature by measuring a change in the voltage. It consists of two different metals: opened and closed. These metals work on the principle of thermo-electric effect. When two dissimilar metals produce a voltage, then a thermal difference exists between the two metals. When the temperature goes up, the output voltage of the thermocouple also increases.

This thermocouple sensor is usually sealed inside a ceramic shield or a metal that protects it from different environments. Some common types of thermocouples include K, J, T, R, E, S, N, and B. The most common type of thermocouples is J, T and K type thermocouples, which are available in pre-made forms.

The most important property of the thermocouple is nonlinearity – the output voltage of the thermocouple is not linear with respect to temperature. Thus, to convert an output voltage to a temperature, it requires mathematical linearization.

Resistor Temperature Detector (RTD)

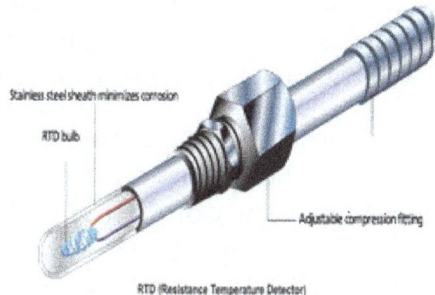

RTD sensor is one of the most accurate sensors. In a resistor temperature detector, the resistance is proportional to the temperature. This sensor is made from platinum, nickel, and copper metals. It has a wide range of temperature measurement capabilities as it can be used to measure temperature in the range between -270oC to +850oC. RTD requires an external current source to function properly. However, the current produces heat in a resistive element causing an error in the temperature measurements. The error is calculated by this formula:

$$\Delta T = P * S$$

Where, 'T' is temperature, 'P' is I squared power produced and 'S' is a degree C/mill watt.

There are different types of techniques to measure temperature by using this RTD. They are two wired, three-wired and four-wired method. In a two-wired method, the current is forced through the RTD to measure the resulting voltage. This method is very simple to connect and implement; and, the main drawback is – the lead resistance is the part of the measurement which leads to erroneous measurement.

Three-wired method is similar to the two-wired method, but the third wire compensates for the lead resistance. In a four-wired method, the current is forced on one set of the wires and the voltage is sensed on the other set of wires. This four-wired method completely compensates for the lead resistance.

Thermistors

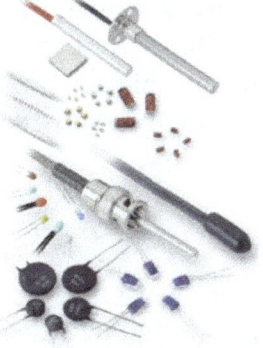

Another type of sensor is a thermistor temperature sensor, which is relatively inexpensive, adaptable, and easy to use. It changes its resistance when the temperature changes like RTD sensor.

Thermistors are made from manganese and oxides of nickel, which make them susceptible to damages. So, these materials are called ceramic materials. This thermistor offers higher sensitivity than the resistor temperature detectors. Most of the thermistors have a negative temperature coefficient. It means, when the temperature increases the resistance decreases.

A thermometer is a device used to measure the temperature of solids, liquids, or gases. The name thermometer is a combination of two words: thermo – means heat, and meter means to measure. Thermometer contains a liquid, which is mercury or alcohol in its glass tube. The volume of the thermometer is linearly proportional to the temperature – when the temperature increases, the volume of the thermometer also increases.

When the liquid is heated it expands inside the narrow tube of the thermometer. This thermometer has a calibrated scale to indicate the temperature. The thermometer has numbers marked alongside the glass tube to indicate the temperature when the line of mercury is at that point. The temperature can be recorded in these scales: Fahrenheit, Kelvin or Celsius. Therefore, it is always desirable to note for which scale the thermometer is calibrated.

Semiconductor Sensors

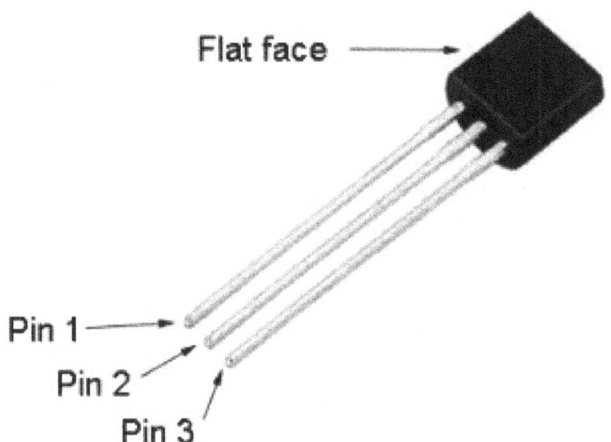

Semiconductor sensors are the devices that come in the form of ICs. Popularly, these sensors are known as an IC temperature sensor. They are classified into different types: Current output temperature sensor, Voltage output temperature sensor, Resistance output silicon temperature sensor, Diode temperature sensors and Digital output temperature sensor. Present semiconductor temperature sensors offer high linearity and high accuracy over an operating range of about 55° C to +150° C. However, AD590 and LM35 temperature sensors are the most popular temperature sensors.

IR Sensor

IR sensor is an electronic instrument which is used to sense certain characteristics of its surroundings by either emitting or detecting IR radiation. These sensors are non-contacting sensors. For example, if you hold an IR sensor in front of your desk without establishing any contact, the sensor detects the temperature of the desk based on the merit of its radiation. These sensors are classified into two types such as thermal infrared sensors and quantum infrared sensors.

Thus, this is all about different types of temperature sensors. The cost of the temperature sensor depends on the type of work it is intended for. However, the accuracy of the sensor will decide the price. So, the cost depends on the accuracy of the temperature sensor.

Application of Temperature Sensor

Design of Industrial Temperature Controller for controlling temperature of devices used in industrial applications is one of the frequently used practical applications of the temperature sensor. In this circuit IC DS1621, a digital thermometer is used as a temperature sensor, thermostat, which provides 9-bit temperature readings. The circuit mainly consists of 8051 microcontroller, EE-PROM, temperature sensor, LCD display and other components.

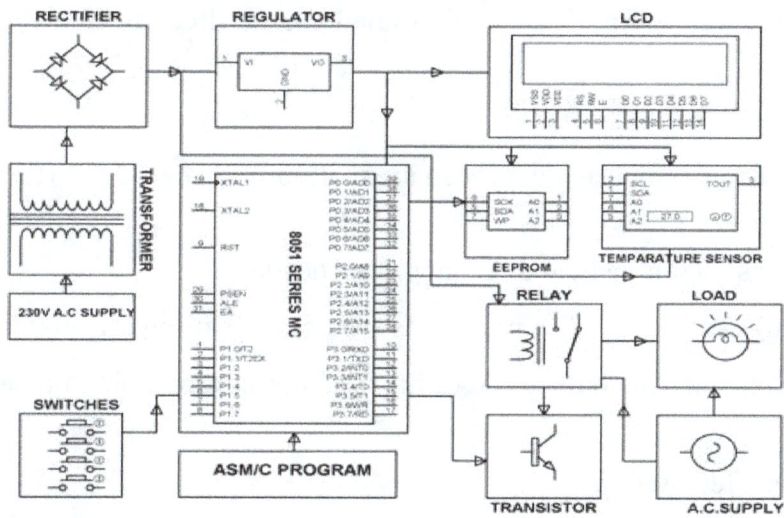

LCD is used to display temperature in the range of -55degress to +125degrees. EEPROM is used to store predefined temperature settings by user through the 8051 series microcontroller. The relay whose contact is used for load, is driven by microcontroller using a transistor driver.

Electronic Sensors

Sensor-based automated control makes lighting simple for people using rooms and spaces. By making smart use of your lighting resource, sensor-based control contributes to meeting building performance and environmental targets.

Sensors are most often used to control lighting depending on room occupancy and levels of ambient light (light already present before any additional lighting is added). This ensures the appropriate level of light is provided at the right time for the expected user experience, and contributes to energy savings.

As part of a lighting control solution, lighting control sensors automatically control lighting using three main methods:

Absence Detection

Where people want to control the lights manually but want to make sure the lights are not left on when a room or space is not in use, absence detection is used.

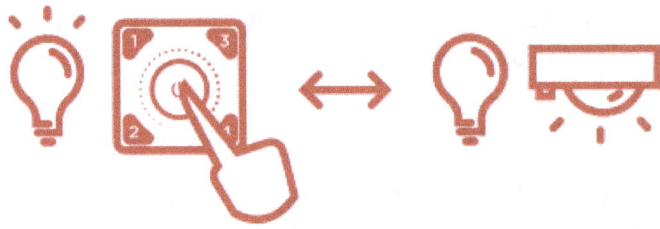

Control process:

1. When someone enters an unlit room, they turn the lights on manually.

2. If nobody is in the room, lights are dimmed then switched off after a pre-determined time.

Benefits:

- Wellbeing and Convenience: People can manually adjust the lights to their particular needs and preferences.

- Energy Savings: Lights switched on only when needed.

- Energy Savings and Peace of Mind: Lights switched off automatically when not needed.

- Intelligent Control: Timeouts and lighting levels programmable to meet the application's requirements.

Absence detection is often used in:

- Classrooms — an example:

    When the teacher arrives for the first class of the morning, they turn the lights on manually. During the lesson, the teacher adjusts the lights to suit different activities, including bright cool white light for a test. After class, the last person out forgets to turn the lights off. 5 minutes later, as the sensor detects that nobody is there, the lights are dimmed low, and two minutes later, the lights are turned off automatically. Twenty minutes later, the teacher returns to the room to collect a book from the front desk. The sensor, set to absence detection, does not turn the lights on. There is enough light from the corridor and windows, so there is no need to switch the classroom lights on again.

- Meeting Rooms — an example:

  The first person to arrive turns on the lights. Shortly after the start of the meeting, a presentation is given. One press of a button dims most lights and lowers the blinds. To begin the question and answer session, the presenter presses a button to turn the lights back up and raise the blinds. Later someone manually adjusts the blinds and the lights.

  As people leave after the meeting, nobody turns the lights off. 8 minutes later, as the sensor has detected no movement in the room, the lights are dimmed, and 3 minutes later, the lights go off, and the control system ensures the blinds are fully raised, to allow in as much daylight as possible.

  With the lights off, people can see, through the window in the meeting room door, that the room is ready and available for the next meeting.

- Offices

- Hotel Rooms.

Presence Detection

Where people want the lights to come on automatically when they enter a room, and want to make sure the lights are turned off when a room or space is not in use, presence detection is used.

Control process:

1. Lights are turned on automatically when someone enters the room.

2. If nobody is in the room, lights are dimmed then turned off after a pre-determined time.

Benefits:

- Wellbeing and Convenience: Lights switched on automatically.

- Safety: Ensure safe levels of lighting in corridors and stairwells whenever someone is there.

- Hygiene and Safety: No need for staff (e.g. medical staff, cleaners, kitchen staff, production workers) to touch button panels.

- Ease of Use: Fully automatic lighting control.

- Energy Savings: Lights switched on only when people in the room.

- Energy Savings and Peace of Mind: Lights turned off automatically when not needed.

- Intelligent Control: Timeouts and lighting levels programmable to meet the application's requirements.

Presence detection is often used in:

- Open-Plan Offices — an example:

  When the first person arrives early in the morning, the office kitchen lights come on automatically, and so do the lights in the main aisle of the open plan office. When they've made themselves a drink, they go to their desk, and the lights come on automatically in all the zones they walk through, up to and including their desk area. As other people arrive at their desks, all the lights in the open-plan office come on. At the end of the working day, one person is still working at the far end of the office. The lights in the zone around their desk are on, and the lights in the main aisle of the office and in all exit corridors are on, so that anyone still in the office has a well-lit work area and feels safe. This is an example of presence detection with multiple sensors working together with a feature called 'corridor hold'.

- Storage Rooms — an example:

  When you open the door to enter the storage room, the lights come on immediately. Without having to find a light switch, it is easy to store or find what you need. Then, when you leave the room, 90 seconds later the lights in the storage room go off automatically. The room is simple to use and you know the lights will not be left on overnight or over the weekend or for longer.

- Corridors

- Stairwells

- Toilets

- Circulation Areas

- Kitchen and Coffee Break Areas.

Constant Light

Modern buildings are expected to maximise use of natural daylight, for people's comfort and wellbeing and to save energy. Daylight Harvesting is often used in rooms and spaces where natural daylight can provide some or all of the light required, and where the electric lights can be adjusted in response to varying ambient light levels.

- Wellbeing and Convenience: Lights switched on automatically

- Safety: Ensure safe levels of lighting in corridors and stairwells whenever someone is there

- Hygiene and Safety: No need for staff (e.g. medical staff, cleaners, kitchen staff, production workers) to touch button panels.

- Ease of use: Fully automatic lighting control

- Energy Savings: Lights switched on only when people in the room

- Energy Savings and Peace of Mind: Lights turned off automatically when not needed

- Intelligent Control: Timeouts and lighting levels programmable to meet the application's requirements.

Presence detection is often used in:

- Open-Plan Offices — an example:

  When the first person arrives early in the morning, the office kitchen lights come on automatically, and so do the lights in the main aisle of the open plan office. When they've made themselves a drink, they go to their desk, and the lights come on automatically in all the zones they walk through, up to and including their desk area. As other people arrive at their desks, all the lights in the open-plan office come on. At the end of the working day, one person is still working at the far end of the office. The lights in the zone around their desk are on, and the lights in the main aisle of the office and in all exit corridors are on, so that anyone still in the office has a well-lit work area and feels safe. This is an example of presence detection with multiple sensors working together with a feature called 'corridor hold'.

- Storage Rooms — an example:

  When you open the door to enter the storage room, the lights come on immediately. Without having to find a light switch, it is easy to store or find what you need. Then, when you leave the room, 90 seconds later the lights in the storage room go off automatically. The room is simple to use and you know the lights will not be left on overnight or over the weekend or for longer.

- Corridors.

- Stairwells.

- Toilets.

- Circulation Areas.

- Kitchen and Coffee Break Areas.

Control Process Step-through

Presence detection (and absence detection) for security checks in an office building:

Presence detection can be implemented with PIR and microwave sensors.

Constant Light

Modern buildings are expected to maximise use of natural daylight, for people's comfort and well-being and to save energy. Daylight Harvesting is often used in rooms and spaces where natural daylight can provide some or all of the light required, and where the electric lights can be adjusted in response to varying ambient light levels.

Helvar's multisensors monitor ambient light levels and via constant light control, adjust electric lighting to maintain the appropriate light level required in the room. In some buildings, the lights and blinds inside are controlled based on readings from a Helvar light sensor outside the building, for example to keep rooms cool in hot sunny locations.

Benefits:

- Wellbeing and Convenience: The required level of light is maintained automatically.

- Ease of use: Fully automatic lighting control to maintain target light level.

- Energy Savings: Electric lights dimmed as natural daylight levels increase.

- Energy Savings: Combine with absence and presence detection to optimise energy savings.

- Intelligent Control: The rate of light adjustment can be precisely set to meet the application's requirements.

- Intelligent Control: Apply different offsets to various groups/zones of lights to maintain similar light levels throughout a room or space.

Constant light is often used in:

- Classrooms — an example:

  When the teacher arrives for the first class of the morning, they turn the lights on manually. It is a sunny day, and the light coming in through the windows is almost enough to light the classroom. The multisensor light level readings enable the lighting control system to dim the electric lights down to a minimum. At lunchtime, the lights are switched off. Late in the afternoon, the lights are switched back on again. As it is quite dark outside, the lighting control system turns the lights up to provide the correct light level for the class, and is it gets darker outside, the lights are gradually turned up to maximum. This change is so gradual that nobody notices. After the lesson, the mulitsensor detects that nobody is there, and the lights are dimmed low, and two minutes later, the lights are turned off automatically.

- Offices

- Hospital Treatment Rooms

- External Linking Corridors.

- Spaces With Skylights or transparet roofs such as in studios and museums.

Relative Humidity Sensors

Resistive Humidity Sensors are humidity sensors that measure the resistance (impedance) or electrical conductivity. The principle behind resistive humidity sensors is the fact that the conductivity in non–metallic conductors is dependent on their water content.

Working of Resistive Humidity Sensors

The Resistive Humidity Sensor is usually made up of materials with relatively low resistivity and this resistivity changes significantly with changes in humidity. The relationship between resistance and humidity is inverse exponential. The low resistivity material is deposited on top of two electrodes.

The electrodes are placed in interdigitized pattern to increase the contact area. The resistivity between the electrodes changes when the top layer absorbs water and this change can be measured with the help of a simple electric circuit.

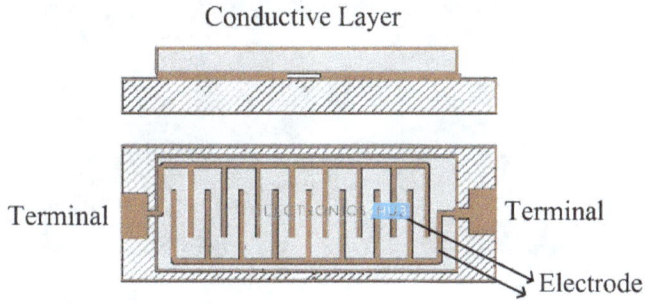

Some of the commonly used materials are salt, specially treated substrates, solid polyelectrolytes and conductive polymers. Modern Resistive Humidity Sensors are coated with ceramic substance to provide extra protection. The electrodes in the sensor are usually made of noble metals like gold, silver or platinum.

Advantages

- Low cost

- Small Size

- The distance between the sensor and signal circuit can be large (suitable for remote operations).

- Highly interchangeable as there are no calibration standards.

Disadvantages

- Resistive Humidity Sensors are sensitive to chemical vapors and other contaminants

- The output readings may shift if used with water soluble products.

Applications

Resistive or Electrical Conductive Humidity sensors are low cost sensors with relatively small size. They are often used in several industrial, domestic or residential and commercial applications.

Pressure Sensors

With the growing requirements of building maintenance, HVAC systems are no longer able to meet demand operating alone. Whether it is a commercial, retail or industrial building, pressure sensors play a significant role in monitoring the building for occupants' safety and comfort. When integrated with pressure technology, HVAC systems can operate at optimum levels as well as improve building maintenance. Three ways by which, this can be done are mentioned below.

Measuring Differential Pressure/Detecting Areas of Maintenance

In a building, quality control is fundamental. Pressure sensors closely monitor pressure changes and indicate if a predicament arises. Pressure drops across rooms or across a filter can alert service technicians for maintenance. Pressure transducers monitoring differential pressure in HVAC systems are especially important for critical applications such as stairwell pressurization, clogged filter detection or cleanrooms.

Areas Used in: Compressors, Coolers, Boilers, Heat Recovery Systems, Burner Control, VAV.

Improving Indoor Air Quality

Sensors make HVAC systems more efficient by measuring air flow and pressure though a system for effective air distribution. Buildings require a certain amount of air flow depending on the building function and average number of occupants. For example since a mall has a large amount of traffic and people emitting CO_2, pressure sensors ensure that a greater amount of airflow is getting pumped back into the building, so individuals can breathe easily. Pressure sensors are especially beneficial in cleanrooms by assessing and monitoring air pressure, quality and ventilation; all important considerations in a cleanroom environment.

Areas Used In: VAV (Variable Air Volume) control, static duct pressure, Air handling systems.

Optimizing Systems

Through measuring pressure and air flow of individual rooms, building owners can optimize heating, cooling, and air flow. By increasing the efficiency of the HVAC system, sensors can reduce the buildings energy consumption and costs. Detecting areas for maintenance can also reduce energy costs by preventing unnecessary energy loads from malfunctioning equipment. Pressure sensors can also ensure compliance with legal required standards.

Pressure-sensing Technology

Front and back of a silicon pressure sensor chip. Note the etched depression in the front; the sensitive area is extremely thin. The back side shows the circuitry, and rectangular contact pads at top and bottom. Size: 4x4 mm.

There are two basic categories of analog pressure sensors:

Force Collector Types

These types of electronic pressure sensors generally use a force collector (such a diaphragm, piston, bourdon tube, or bellows) to measure strain (or deflection) due to applied force over an area (pressure).

- *Piezoresistive strain gauge*

Uses the piezoresistive effect of bonded or formed strain gauges to detect strain due to applied pressure, resistance increasing as pressure deforms the material. Common technology types are Silicon (Monocrystalline), Polysilicon Thin Film, Bonded Metal Foil, Thick Film, Silicon-on-Sapphire and Sputtered Thin Film. Generally, the strain gauges are connected to form a Wheatstone bridge circuit to maximize the output of the sensor and to reduce sensitivity to errors. This is the most commonly employed sensing technology for general purpose pressure measurement.

- *Capacitive*

Uses a diaphragm and pressure cavity to create a variable capacitor to detect strain due to applied pressure, capacitance decreasing as pressure deforms the diaphragm. Common technologies use metal, ceramic, and silicon diaphragms.

- *Electromagnetic*

Measures the displacement of a diaphragm by means of changes in inductance (reluctance), LVDT, Hall Effect, or by eddy current principle.

- *Piezoelectric*

Uses the piezoelectric effect in certain materials such as quartz to measure the strain upon the sensing mechanism due to pressure. This technology is commonly employed for the measurement of highly dynamic pressures.

- *Strain-gauge*

Strain gauge based pressure sensors also use a pressure sensitive element where metal strain gauges are glued on or thin film gauges are applied on by sputtering. This measuring element can either be a diaphragm or for metal foil gauges measuring bodies in can-type can also be used. The big advantages of this monolithic can-type design are an improved rigidity and the capability to measure highest pressures of up to 15,000 bar. The electrical connection is normally done via a Wheatstone bridge which allows for a good amplification of the signal and precise and constant measuring results.

- *Optical*

Techniques include the use of the physical change of an optical fiber to detect strain due to applied pressure. A common example of this type utilizes Fiber Bragg Gratings. This technology is employed in challenging applications where the measurement may be highly remote, under high temperature, or may benefit from technologies inherently immune to electromagnetic interference. Another analogous technique utilizes an elastic film constructed in layers that can change reflected wavelengths according to the applied pressure (strain).

- *Potentiometric*

Uses the motion of a wiper along a resistive mechanism to detect the strain caused by applied pressure.

Other Types

These types of electronic pressure sensors use other properties (such as density) to infer pressure of a gas, or liquid.

- *Resonant*

Uses the changes in resonant frequency in a sensing mechanism to measure stress, or changes in gas density, caused by applied pressure. This technology may be used in conjunction with a force collector, such as those in the category above. Alternatively, resonant technology may be employed by exposing the resonating element itself to the media, whereby the resonant frequency is dependent upon the density of the media. Sensors have been made out of vibrating wire, vibrating cylinders, quartz, and silicon MEMS. Generally, this technology is considered to provide very stable readings over time.

- *Thermal*

Uses the changes in thermal conductivity of a gas due to density changes to measure pressure. A common example of this type is the Pirani gauge.

- *Ionization*

Measures the flow of charged gas particles (ions) which varies due to density changes to measure pressure. Common examples are the Hot and Cold Cathode gauges.

Applications

There are many applications for pressure sensors:

- Pressure sensing

This is where the measurement of interest is pressure, expressed as a force per unit area. This is useful in weather instrumentation, aircraft, automobiles, and any other machinery that has pressure functionality implemented.

- Altitude sensing

This is useful in aircraft, rockets, satellites, weather balloons, and many other applications. All these applications make use of the relationship between changes in pressure relative to the altitude. This relationship is governed by the following equation:

$$h = (1 - (P/P_{ref})^{0.190284}) \times 145366.45 \text{ft}$$

This equation is calibrated for an altimeter, up to 36,090 feet (11,000 m). Outside that range, an error will be introduced which can be calculated differently for each different pressure sensor. These error calculations will factor in the error introduced by the change in temperature as we go up.

Barometric pressure sensors can have an altitude resolution of less than 1 meter, which is significantly better than GPS systems (about 20 meters altitude resolution). In navigation applications altimeters are used to distinguish between stacked road levels for car navigation and floor levels in buildings for pedestrian navigation.

- Flow sensing

This is the use of pressure sensors in conjunction with the venturi effect to measure flow. Differential pressure is measured between two segments of a venturi tube that have a different aperture. The pressure difference between the two segments is directly proportional to the flow rate through the venturi tube. A low pressure sensor is almost always required as the pressure difference is relatively small.

- Level/depth sensing

A pressure sensor may also be used to calculate the level of a fluid. This technique is commonly employed to measure the depth of a submerged body (such as a diver or submarine), or level of contents in a tank (such as in a water tower). For most practical purposes, fluid level is directly proportional to pressure. In the case of fresh water where the contents are under atmospheric pressure, 1psi = 27.7 inH20/1Pa = 9.81 mmH20. The basic equation for such a measurement is,

$$P = \rho g h$$

where P = pressure, ρ = density of the fluid, g = standard gravity, h = height of fluid column above pressure sensor.

- Leak testing

A pressure sensor may be used to sense the decay of pressure due to a system leak. This is commonly done by either comparison to a known leak using differential pressure, or by means of utilizing the pressure sensor to measure pressure change over time.

Ratiometric Correction of Transducer Output

Piezoresistive transducers configured as Wheatstone bridges often exhibit ratiometric behavior with respect not only to the measured pressure, but also the transducer supply voltage.

$$V_{out} = \frac{P \times K \times Vs_{actual}}{Vs_{ideal}}$$

where:

V_{out} is the output voltage of the transducer.

P is the actual measured pressure.

K is the nominal transducer scale factor (given an ideal transducer supply voltage) in units of voltage per pressure.

Vs_{actual} is the actual transducer supply voltage.

Vs_{ideal} is the ideal transducer supply voltage.

Correcting measurements from transducers exhibiting this behavior requires measuring the actual transducer supply voltage as well as the output voltage and applying the inverse transform of this behavior to the output signal:

$$\frac{V_{out} \times Vs_{ideal}}{K \times Vs_{actual}}$$

Common mode signals often present in transducers configured as Wheatstone bridges are not considered in this analysis.

Flow Sensors

A Flow Sensor that flow of liquid media in pipes, can be used mostly for water shortage protection in HVAC systems. The monitor is equipped with a potential-free changeover switch responsible for reliably activating an actuator. The second sensor measures air flow that controls non-aggressive gaseous flows in the range of 0.5 to 10/30 m/s. Available in both 0-10V or 4-20mA output. Additionally, a separate 0-10V output signal for temperature pick-up can be added.

A flow sensor, the sensor surface is heated to a defined temperature. If this surface is exposed to a flow, the Senor surface cools. A greater flow leads to a greater cooling and serves as a measure for the flow speed. The cooling effect of the flow is electronically evaluated and the flow sensor supplies depending on model a hereupon based output signal, which can be used for flow monitoring. The applications of flow sensors are diverse, ranging from the general air conditioning systems, exhaust systems, to monitoring of cooling and lubricant circuits. In contrast to the indirect flow control by controlling the rotational movement of a fan or pump, is on this principle directly controls the flow of the medium and thus enables the detection of dirty filters and blockages or blockages in the flow circuit. The adjustable flow value can be an early warning offalling below a minimum flow supply and so z. B. initiate timely maintenance of air conditioning. Due to an absolutelymaintenance-free and wear-free operation Proxitron flow sensors in comparison with other sensor systems such as wind vanes and vane anemometers relays are superior and are preferably used where safe and trouble-free flow control is required.

The construction of the sensors in a seamless closed housing predestined these particularly suitable for use in environments with high contamination levels. Proxitron Air flow sensors (air flow monitor) in solid plastic housing are everywhere at home, where aggressive chemical atmospheres prevail. They resist even containing sulfuric exhaust of electroplating bathswork and many years of trouble free. flow sensors Proxitron are available in various designs. In principle, between sensors formonitoring gaseous and liquid media distinguished.

Flow sensors for gases (also called air flow monitor) are in most cases provided with a compact plastic housing that the easy installation allows in existing exhaust ventilation or ventilation systems. The Flow sensors for liquids have stainless steel process connections in the common pipe thread sizes. Available in both cases variants for AC or DC operating voltage, with different switching output types or with analog output. The units are delivered with a fixed connection cables of different lengths or with conventional M12 connector. Proxitron flow sensors are also characterized by their ease of use from. On push of a button, the sensor will detect the existing flow and represents the flow limit so that a stall is detected. This has diverse applications in different industries, but also in the HVAC and building automation. A matching accessories completes the program and also allows for the flameproof installation of the sensors.

References

- Tech-know-pneumatic-control-systems: acs-southeast.com, Retrieved 28 May 2018
- 6-different-types-of-temperature-sensors-with-their-specifications: edgefx.in, Retrieved 30 March 2018
- Different-types-of-sensors-with-applications: edgefx.in, Retrieved 18 June 2018
- Lighting-control-sensor-concepts: helvar.com, Retrieved 28 April 2018
- Humidity-sensor-types-working-principle: electronicshub.org, Retrieved 15 May 2018
- Why-do-hvac-systems-need-pressure-sensors: setra.com, Retrieved 26 June 2018

Permissions

Index